The Hypersensitive Reaction in Plants to Pathogens

A Resistance Phenomenon

R. N. Goodman
Emeritus Professor

and

A. J. Novacky
Professor

Department of Plant Pathology
University of Missouri
Columbia

APS PRESS
The American Phytopathological Society
St. Paul, Minnesota

Cover: Tobacco leaf tissue exhibiting HR-induced general membrane damage 6 h after inoculation with *Pseudomonas syringae* pv. *pisi.* (Courtesy R. N. Goodman)

This book has been reproduced directly from computer-generated copy submitted in final form to APS Press by the authors. No editing or proofreading has been done by the Press.

Reference in this publication to a trademark, proprietary product, or company name by personnel of the U.S. Department of Agriculture or anyone else is intended for explicit description only and does not imply approval or recommendation to the exclusion of others that may be suitable.

Library of Congress Catalog Card Number: 94-70285
International Standard Book Number: 0-89054-165-5

Second printing, 1996

Printed in the United States of America on acid-free paper

The American Phytopathological Society
3340 Pilot Knob Road
St. Paul, Minnesota 55121-2097, USA

TABLE OF CONTENTS

Acknowledgements

The authors are deeply indebted to a number of colleagues for their assistance in providing data and commentary from the historical to the most current. Their willingness to provide information not yet published has permitted us to bring to the reader the very latest developments in this rapidly moving field of plant science.

However, every effort such as this one clearly reflects the tireless input and devotion to detail concerned with veracity of citations and above all coherent language, punctuation, order and general uniformity. In our instance this has been the role, and much more, of Ms. Sharon Pike, research associate to Professor Novacky. The role of interpreter of handwriting scrawl of the authors and their disjuncted word processor copy necessitating endless correction and rearrangement was handled, without loss of cool, by our secretary, Ms. Marlene Schofer. The diligent word for word editing by Dr. Tibor Ersek, who assumed the role of silent arbiter, overrode the oft-divergent copy of the authors. Review of copy during various stages of development by the experts; Professors Cornelia and Wolfram Ullrich, Margaret Essenberg, Tibor Ersek, Guri Johl, Kohei Tomiyama, Joseph Kuć, Marvin Weintraub, Noel Keen and Milton Zaitlin is also deeply appreciated. The authors are fully aware of the importance that the Rockefeller Study Center, Villa Serbelloni, Bellagio, Italy, played in the development of the early stages of this book. Our treatment of HR had its genesis in the magnificent atmosphere of the Center and its total support to creativity. We thank Rockefeller, the late Roberto Celli, Director, Gianna Celli and her staff for their attention to all of our needs.

Foreword

Both authors have for more than two decades conducted research on a general defense system in plants against pathogens. Our efforts have focused primarily on the induction of this defense system, the hypersensitive reaction (HR), by plant pathogenic bacteria. Nevertheless, it is clear that HR is operative in plants infected by viruses and fungi as well.

We propose to describe this phenomenon as it develops across the broad array of host-pathogen interactions. The historical aspects, biochemical-physiological, ultrastructural and specific genetic factors implicated in the induction and development of HR are defined in some detail. We summarize our understanding of how HR proceeds, based primarily on data from research conducted on bacterial elicitation of the phenomenon. Finally, a number of unsolved basic questions pertinent to the induction and development of HR remain, and these will be delineated.

The authors dedicate this book to a few of the many who, over the years, have studied HR most intensively. In this regard we call to the reader's attention the efforts of Professors Kohei Tomiyama, Hokkaido University, Sapporo, Japan, Professor Joseph K. Kuć, University of Kentucky, Professor Zoltan Klement, Budapest, Hungary, and Dr. Marvin Weintraub, Agriculture Canada, BC.

The Hypersensitive Reaction in Plants to Pathogens

A Resistance Phenomenon

"...work...was undertaken in order to determine whether the phenomenon (HR) was one of real resistance or an extreme case of hyper-sensitiveness".... Stakman 1915.
The question may now be answered as ***real resistance*** *..... The authors 1994.*

THE FUNGUS-INDUCED HYPERSENSITIVE REACTION (HR)

Some historical aspects

The earliest studies pertinent to HR were directed at understanding the resistance of plants to some obligate fungi. Gäumann (1950), in describing a seminal study by Allen (1926) on the infection of the hypersensitive wheat cv. Malakoff by *Puccinia graminis tritici*, wrote that when the fungus sends "haustoria" into the mesophyll cells, the latter often react with utmost intolerance, they degenerate and die. Gäumann continues by noting that "injurious as this over-sensitivity is for the hyperergically reacting cells, it is yet very favorable to the plant as a whole since the focus of the disease is localized and the parasite is prevented from reproducing itself". The terminology of Gäumann becomes even more creative in the subsequent discussion of some details of HR. Two varieties of wheat (compatible-Little Club and incompatible Malakoff) reacted to *Puccinia* similarly according to Gäumann. "Both varieties are susceptible to infection by the pathogen and, in both, the plasmatic defense reactions are inadequate to destroy the fungus during the *process of penetration.* The two varieties of wheat differ only in their secondary behavior. The compatible, normally susceptible, variety tolerates the parasite whereas the incompatible, hypersensitive variety does not tolerate the parasite, it reacts idiosyncratically to its secretions ... thereby hinders or stops the parasite and the plant develops only small local foci of disease."

Stakman (1915) is generally credited with the use of the term, *hypersensitive reaction*, to denote a plant response to inoculation with a fungus and the early development of the pathogen in host tissue under these circumstances. In his account of the comparative infection processes of *Puccinia* species in oats, rye, wheat and barley, Stakman notes that some "biologic forms or strains" killed host cells shortly after inoculation in cultivars that exhibited a considerable degree of resistance (Fig. 1). He wrote that "unquestionably the host plant in such cases is often *hypersensitive* to the fungus". Stakman clearly equated *hypersensitiveness* directly with "the abnormally rapid death of host plant cells attacked by rust hyphae".

Hypersensitive cell death was preceded by chloroplast degeneration and subsequent total loss of cytoplasmic structure. Interestingly, "point" destruction of chloroplasts on one side of a host cell was detected in response to closely appressed fungal hyphae. Histological observations of "normal" infections revealed that inoculated host cells remained apparently healthy for a considerable length of time. Neither extensive nor rapid killing of tissues took place. The time course recorded for symptoms of HR in resistant tissues and abundant spore production in susceptible tissue is 2-3.5 days for the former and 7-12 days for the latter.

The early and important observations by Ward (1902) and by Stakman (1915) were summarized by Gäumann (1950); the fungus enters both susceptible and resistant hosts in the same way but develops much differently thereafter. In the susceptible host the fungus grows rapidly, yet without appearing to affect host cells for some time. However, in resistant hosts a rapid reaction develops that results in "the almost immediate death of some host cells" (Stakman 1915). *Stakman observed that the more resistant a cultivar, the more rapid the death of a limited number of cells in the vicinity of the invading hyphae.*

A *raison d'etre* for this phenomenon was offered earlier by Marryat (1907): the possible starvation of the pathogen resulting from the death of the affected host cells. Stakman seemed unconvinced of this explanation, sensing "a very definite antagonism between the *immune plant and the parasite*". The question concerning the mechanism of the localized rapid host cell death is still not fully resolved at this writing.

The concept of a causal relationship between hypersensitivity (rapid host cell death) and resistance was developed, with a degree of concensus, in a sequence of research

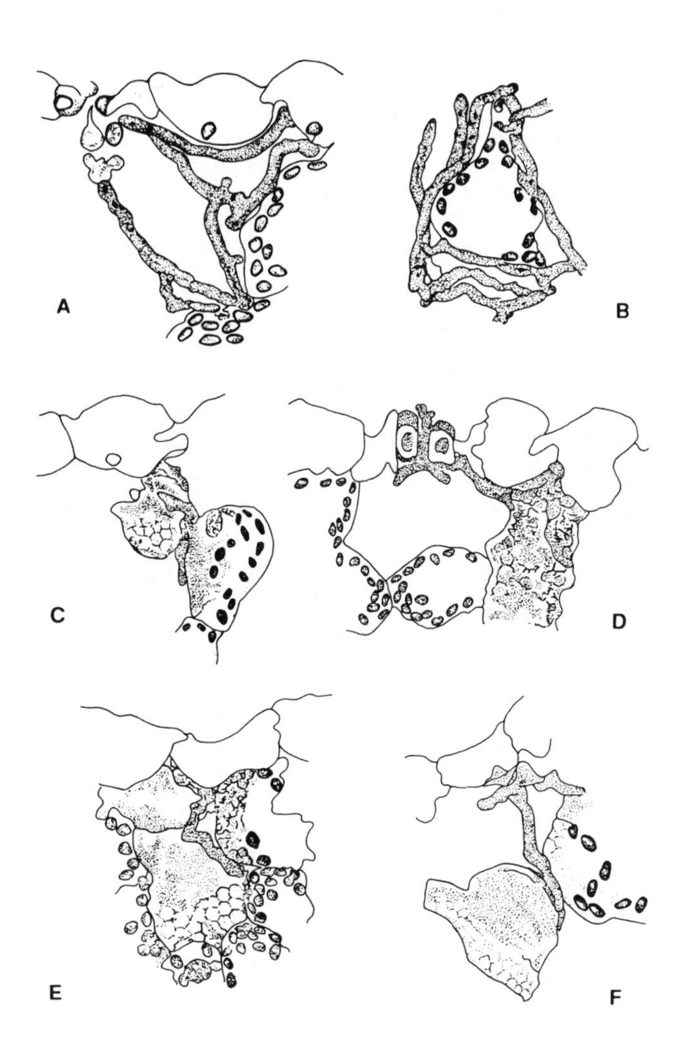

Fig. 1. The outlines in this illustration were made with a camera lucida and the detail was studied under a magnification of 1,000.

a. - Oats inoculated with *Puccinia graminis* from *Dactylis glomerata* after three generations on oats. Early infection stage, showing haustoria in the two epidermal cells to the right of the stoma and one in mesophyll cell. Host cells normal; infection normal and successful.

b. - Same as figure 1a, seven days after inoculation. Cell from edge of pustule area surrounded by hyphae, but still normal.

c. - Oats inoculated with *P. graminis hordei*, four days after inoculation. Piece of mycelium growing over cell; chloroplasts being destroyed; cell at right just being affected; haustorium in epidermal cell.

d. - Oats inoculated with *P. graminis tritici* four days after inoculation. Hypha growing over two cells, both of which have been killed; outlines of chloroplasts still showing faintly in second cell; cell to right of hypha becoming affected.

e. - Oats inoculated with *P. graminis tritici*, 48 hours after inoculation. The cell on the left killed; outlines of chloroplasts still showing very faintly; cell on the right just becoming affected; tip of hypha dying.

f. - Oats inoculated with *P. graminis tritici*, 48 hours after inoculation. The cell on the left killed; outlines of cheoroplasts still showing very faintly; cell on the right just becoming affected; tip of hypha dying. Stakman, 1915.

papers, reviews and books by Ward (1902), Stakman (1915), Arthur (1929), Gäumann (1950), Barnett (1959), and Müller (1959). However, there also have been conflicting perspectives, Brown *et al.* (1966) presented data, and called attention to observations by others (Allen, 1926), that infection in resistant hosts is not always associated with the development of extensive necrotic tissue. It is of more than passing interest that Ward (1902) also mentioned a condition of incompatibility in which no haustoria are formed, even though germination and penetration occur. This is clearly a form of resistance wherein no necrosis develops. Although it was initially inferred that HR resulted in localization and death of the pathogen in incompatible interactions, excision of hyphae from the HR milieu revealed them to be pathogenically competent in susceptible tissue (Sharp and Emge, 1958). By analogy, the change of an apparently resistant reaction to a susceptible one by raising the incubation temperature suggested that hypersensitive necrosis does not kill the fungus (Silverman, 1959; Zimmer and Schafer, 1961). Thus Brown *et al.* (I966) concluded that necrotic tissue in resistant hosts may merely reveal those hosts that are "more sensitive" to the disturbances caused by the invading fungus and that necrosis may be the consequence rather than the cause of resistance. This contention has been held by others and, with respect to the essentiality of necrosis to HR, may be valid.

As late as the early 1960's, little definitive evidence for a causal relationship between HR and resistance had been put forth for plant disease caused by fungi. Indeed Brown *et al.* (1966) emphasized that infection in resistant hosts is not *always* associated with the appearance of extensively necrotized tissue. Thus the resistance conferred by the *Sr6* and the *Sr9* genes of wheat to *P. graminis tritici* may develop by a mechanism other than HR. However, the studies by Allen (1923) reveal that in wheat inoculated with *P. graminis tritici*, highly resistant varieties react in a decidedly hypersensitive manner. Germ tubes from the uredospores penetrate the stomata and develop substomatal vesicles which contact host cell surfaces. The tips of the hyphae form haustorium mother cells, which in turn penetrate host cell walls and invaginate the protoplasts. Rapidly, a series of degenerative changes occur in the haustorium mother cell, the penetrating appendage and the host cell itself, which quickly leads to the death of all three. Subsequent haustorial branches repeat the process leading to the death of a number of neighboring host cells. As a consequence the pathogenic attack is localized and in some instances the pathogen dies. It is clear that in the infinite numbers of possible host-pathogen interactions there is also latitude in the degrees of intensity of HR and susceptibility *as well as other forms of resistance.*

The literature recording HR for *Puccinia* has parallels in observations of diseases caused by a large number of obligate parasites (*i.e.* downy and powdery mildews) (White and Baker, 1954). There is significant evidence that HR, when equated with local lesion formation, also occurs in a number of interactions between plants and facultative fungi. Wood (1967) suggested that HR is apparently a general phenomenon that develops in plants infected by obligate, "intermediate" types such as *Phytophthora infestans* and the facultative fungi. The essential feature in all three is that infection of the host is followed by the rapid death of both the invaded and some neighboring cells. Wood calls our attention to the fact that intense limited host cell death, following infection, may be readily characterized as HR. However, when the process is slower and involves larger areas of host tissue, the phenomenon borders on susceptibility. Perhaps in these latter instances *other* mechanisms are responsible for the ultimate restriction of the pathogen. However, the possibility remains that the same reactions occur in both HR and slow necrotization. The differences perhaps reflect only the rate and scope of cell death.

HR - defined

Carrying Wood's observation a step further it is possible for us to achieve consensus concerning what is or may not be HR by examining the parameters initially set by Stakman (1915) and the observations by Ward (1902). Specifically, HR involves the extremely rapid death of only a few host cells, which limits the progression of the infection. Hence, the criteria are the rapid rate of reaction, the limited numbers of dead cells involved, and the fact that the disease is arrested. It would seem that neither *extensive necrosis* (visible to the unaided eye) nor *death of all hyphae* are imperatives.

Clearly the hypersensitive reaction, with its characteristic rapid cell death and subsequent necrosis, constitutes one of the primary mechanisms of resistance to plant disease. HR varies from other plant disease resistance mechanisms; ***in fact it is the antithesis of the others***. Whereas other resistance mechanisms such as the development of papillae, callose, phytoalexins, cuticle, pellicle, suberin and lignin all involve the synthesis of a protective cell-sparing entity, HR, on the other hand, requires rapid host cell death.

Hence, the entire process of rapid host cell death, its every detail, must be fully understood. In the view of the authors, some previous definitions of HR have been too inclusive, accommodating events that mediate plant disease resistance activities prior or subsequent to or in the absence of host cell death. In our view HR is a reflection of host cell membrane dysfunction that quickly manifests itself in death of the infected cells and *necrosis is but a consequence of cell death*. Localization of the pathogen and its subsequent death or suppression are distinct from those events that foster HR.

The fungus model

Since plant cell death is the central phenomenon of HR, it is necessary to define it in a temporal sense. Although HR has been demonstrated as a defense mechanism for both necrotrophic and biotrophic fungi, a visual, morphologic, physiologic and biochemical sense of the process has been most completely developed for the interaction of resistant potato tissue with *Phytophthora infestans.* In this regard the efforts of Professor Kohei Tomiyama, his students and associates have been legion. As we shall see, for almost 40 years, data concerning this host-pathogen interaction have come from his laboratory and now from his former students and associates. It is perhaps prophetic that this host-pathogen combination should have become, at least to some extent for fungi, the model system.

The announcement in *Nature* in 1949 by K.O. Müller and his associate Luther Behr described rather fully the resistance inherent in the R_1 potato gene to some races of *P. infestans.*

Specifically, they reported the rapidity of necrosis in the two extreme instances: in the susceptible tuber "the parasite penetrating through the entire tuber and fructifying luxuriantly" and "in the highly resistant the parasite penetrating only a few cell layers of the parenchyma of the tuber and then ceasing to grow without fructifying". Müller and Behr observed that the R gene confers the capacity of acquiring a local immunity from infection when potato cells come in contact with the hyphae of "parabiontic" races of *P. infestans*. It was implied that those races which kill plant cells quickly, thereby limit further growth of the fungus or subsequently invading fungi. They further characterize the reaction as proceeding similarly in susceptible and resistant tissues. In both, the host cell is destroyed after coming into contact with the protoplasm of the parasite. Morphological and physiological changes are the same; the cell collapses and loses its ability to serve the parasite as host. The difference between the two is, however, apparent from the rapidity with which the effect is achieved in resistant cells. While the cells without the R gene survive 6-14 days at 19-21° C, the resistant cells die after one or two days. As we shall see from the studies of Tomiyama (1956), some infected cells die within minutes after the host cell wall has been breached.

Müller and Behr also describe an additional consequence of the "defence necrosis". In the course of the reaction, host tissue loses not only its ability to serve as substrate to *P. infestans,* but in addition, becomes refractory to a number of other fungal species. This subsequent effect is due to the *de novo* synthesis of what Müller and Behr refer to as *phytoalexins.*

Of additional interest was the capacity of low MW n-alcohols to narcotize the necrotic process and the subsequent synthesis of phlobaphene (phytoalexin), thus keeping the host cells alive. As a result, mycelial development was maintained in the host and a sequence of specific, gene-controlled metabolic processes was suppressed.

Müller and Behr concluded from this research that the observed necrobiosis is in some way linked to the respiratory metabolism of the host cell. They also contended that the more rapid the reaction by host cells in contact with the fungus to the metabolic products excreted (elicitor) by the fungus, the more resistant the plant.

In the Müller and Behr report we are shown two defense reactions, HR and phytoalexin synthesis. These are probably linked, but are separable reactions. An analysis of some data that permits partitioning of these two reactions will be presented subsequently (Sato and Tomyama, 1976a).

Histological chronology of HR - induced cell death

In the early phases of research by the Tomiyama school, the model system was either cut surfaces of potato tuber disks or midrib cells of potato leaves inoculated with either a virulent or avirulent race of the fungus. Additionally, resistant and susceptible cultivars were inoculated with a single virulent race of the pathogen.

Vital staining of the inoculated tissue with neutral red under conditions that fostered plasmolysis, made it possible to assess microscopically when death occurred, on the average, in the cell population being observed. Kitazawa *et al.* (1973) showed that penetration of both the *susceptible* and *resistant* plant cell walls occurred at the same time, approximately 69 - 100 min after inoculation. Shortly after penetration and the formation of the primary intracellular mycelium, "remarkable" changes were noted in the resistant cultivar. In 10-60 min the first signs of injury were visible: cytoplasmic streaming accelerated, granules that moved with a Brownian motion appeared in the cytoplasm in the vicinity of the point of penetration, and finally cyclosis ceased. After an additional 10-60 min the contents of the infected cell began to discolor. Within 10-30 min of the onset of discoloration, vibratory movement of the granules ceased and the cell contents seemed to "gelatinize" and the plant cell died. Hence, within about 30-150 min after host cell wall penetration in an incompatible interaction, HR-death of the first infected cells was revealed. Necrosis continued for another 10 hr after cell death as the cells became brown and eventually blackish-brown in color (Tomiyama, 1956a,b; Kitazawa and Tomiyama, 1969), see also Fig. 2.

In the infected midrib cells of the *susceptible* variety cytoplasmic Brownian motion did not become apparent until 2-8 h after penetration. During the subsequent 7-10 h no cytoplasmic discoloration developed and approximately 20% of the infected cells appeared unaffected for about 24 h. Clearly, in the compatible interaction the fungus was unable to elicit the HR in the infected cells and the pathogen proceeded to develop unimpeded.

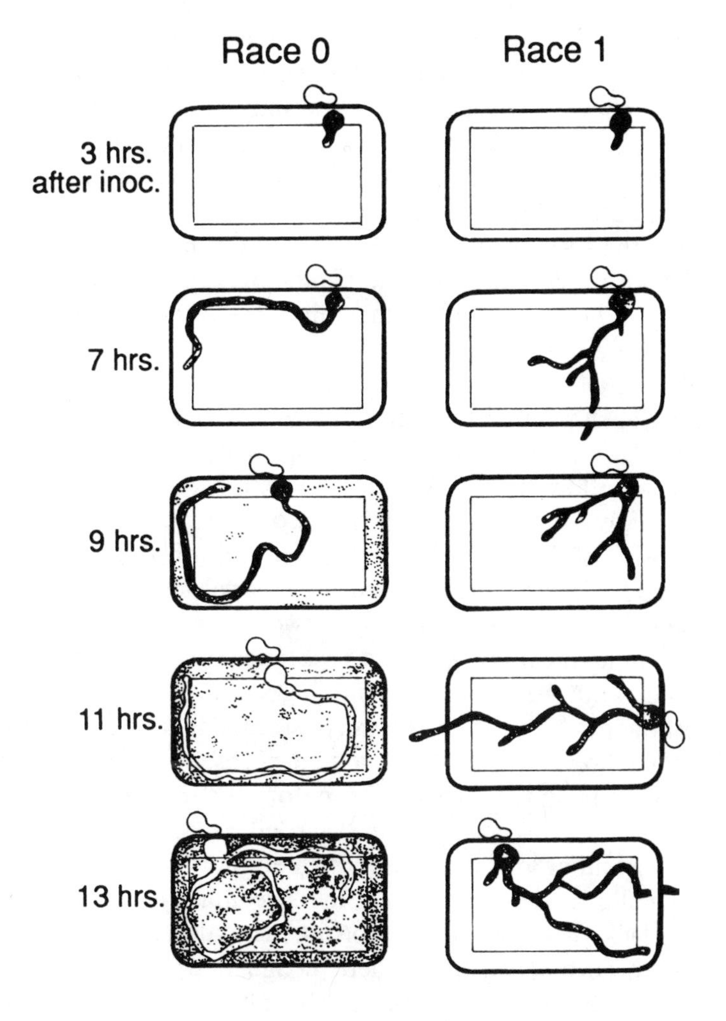

Fig. 2. A sequence of invasion of host cells by the incompatible (left) and compatible (right) race of *P. infestan.* Kitazawa *et al.*, 1973.

It was also noted at this time that the epidermal midrib cells of older tissues took longer to die. This apparently suggested to Tomiyama that the metabolically active cells were able to respond hypersensitively more quickly than less active ones. In a later study Tomiyama (1960) used cut cortical cells of leaf petioles that were aged for 16-20 h prior to inoculation as his model to monitor HR cell death. This treatment greatly elevated HR potential as some cortical cells died in 10-20 min after penetration (Tomiyama, 1967). His observations suggested significant variability in the inoculated cell population with respect to the time at which cell death occurred. Microscopic observations exposed a correlation between the length of intracellular hyphal growth and the time required for cortical cell death. It was found that 10% of the inoculated host cells in an incompatible interaction died within minutes after penetration when the intracellular hyphae were 2 μm in length. About 30% died within 20 min when the hyphae were about 10 μm long and the rest died within 40-60 min when the intracellular hyphae had reached a length of 14 μm.

Relationship of cyclosis suppression to HR

Does the cessation of cytoplasmic streaming characterize HR in other host-pathogen interactions? The answer is the affirmative. Bushnell (1981) also observed the cessation of cyclosis as the first detectable feature of dying exhibited by resistant barley cells infected with barley mildew. Cytoplasmic streaming was monitored in two near isogenic lines, one containing the *mla*R gene for resistance to *Erysiphe graminis* f. sp. *hordei* and the other the *mla*S gene for susceptibility to the pathogen. A precise cytoplasmic streaming index, consisting of 6 criteria, revealed that movement ceased about 27.5 h after inoculation and on average 2 h before the infected cells collapsed (Fig. 3 and 4). Cells were regarded as being in the quiescent stage (neither long distance movement nor localized movement of organelles is detected) when the streaming index fell below 1.0. The lack of Brownian motion in the quiescent stage suggested that the cytoplasm had gelled (coagulated), the conclusion also reached by Tomiyama and his associates (1956a, 1971). Clearly, the cessation of cytoplasmic movement was shown to be a consistent event preceding HR collapse of host tissue. Bushnell (1981) also observed that cell collapse occurred only after the haustorium had partly formed and was clearly visible in the infected cell. Additionally, one or more cells adjacent to a haustorium-containing cell also frequently died. Although cytoplasmic streaming ceased as early as 1 h after host wall penetration in the *P. infestans* model, the sequence of events in the HR in barley and potato, were regarded by Bushnell as similar. Indeed Bushnell interpreted his data as suggesting that the cessation of cytoplasmic streaming "is a general event preceding hypersensitive cell death in gene-for-gene incompatibility". He also inferred that the resultant cell collapse and death reflected a restriction in the ATP supply through interference with the oxidative metabolism of the cell. The latter is a recurring theme throughout many studies of HR on a variety of hosts and pathogens.

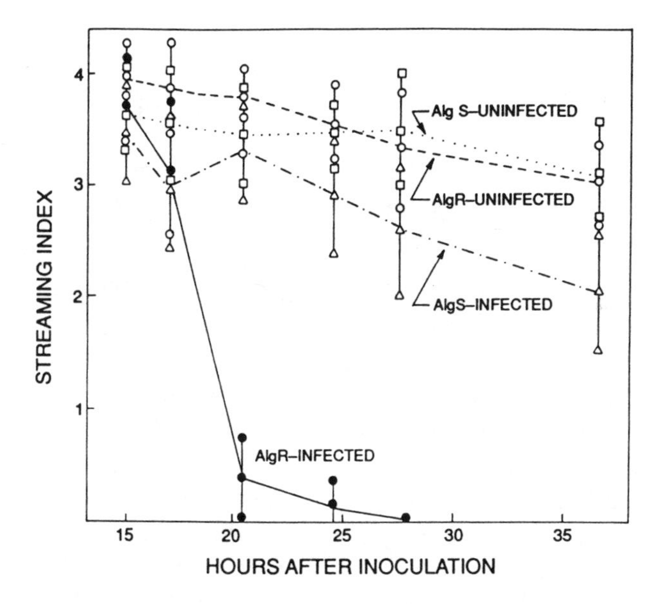

Fig. 3. The cytoplasmic streaming index in barley epidermal tissues uninfected and infected with *Erysiphe graminis* f. sp. *hordei*. Near-isogenic lines, AlgS (compatible) and AlgR (incompatible) differ at the *mla* locus for compatibility with the fungus. Bushnell, 1981.

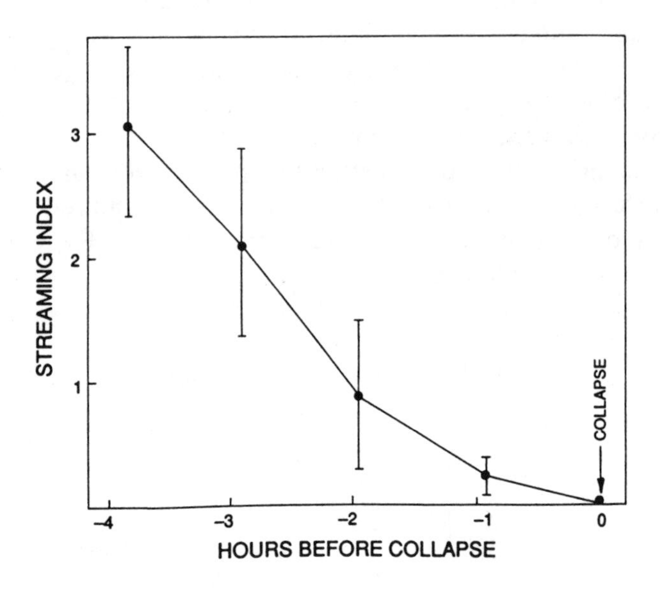

Fig. 4. Streaming index 0-4 h before hypersensitive cell collapse in incompatible epidermis tissues of AlgR under attack by *Erysiphe graminis*. Bushnell, 1981

The experiments of Meyer and Heath (1988a, b) also tend to support the observations of Tomiyama. Meyer and Heath explored the possibility of mimicking the sequence of cowpea cell death caused by *Erysiphe cichoracearum* with cellular death caused by metal salts such as $CuCl_2$, $CuSO_4$ and $HgCl_2$. With the light microscope they observed that cell degeneration and death proceeded in three stages: (1) the slowing and eventual cessation of protoplasmic streaming (2) the formation of large vesicles containing granules and (3) protoplast collapse (cell death). It was observed that semipermeability of the tonoplast was lost in stage 2 and that of the plasmalemma in stage 3. Whether such loss in individual cells occurred in the order of tonoplast first, followed by the plasmalemma was not established. However, in studies by Goodman and Plurad (1971) and Goodman, Huang and White (1976), ultrastructural evidence for the early destruction of the major vacuole in tobacco leaf parenchyma tissue 3-4 h after inoculation with HR-inducing *Pseudomonas syringae* pv. *pisi* suggests that the tonoplast may be particularly fragile. Nevertheless, the similarity between the observations by Meyer and Heath (1988), Bushnell (1981), and Tomiyama (1956b) suggests that changes in protoplasmic streaming are early physiological alterations in plant cells undergoing HR-related death and necrosis.

Ultrastructural and histological aspects of HR induced by biotrophic fungi

It has long been evident that HR reflects disturbance of the host cell's vital membranes. As a consequence a number of investigators have sought the target sites on the membraneous organelles of the cell. Which membranes were attacked first? Would it be possible to actually see early stages of membrane disorientation? Could the process of decompartmentalization be discerned? Or could one relate the presence of the fungus to initial stages of ultrastructural change?

As described in the previous section, Meyer and Heath (1988) induced death in cowpea leaf epidermal cells with heavy metal salts and followed the process with both light and electron microscopy. Light microscopy revealed that death followed the sequence of a) the cessation of cytoplasmic streaming, b) formation of large vesicles in the cytoplasm and c) protoplast collapse. The changes inherent in the cessation of cytoplasmic streaming noted light microscopically correlated well with the observed ultrastructural changes.

First microtubules appeared to decrease in number; then coiled polyribosomes, endoplasmic reticulum and Golgi bodies rapidly became undetectable. These are distinct signs of a rapid loss of synthesizing capacity. The dilation of mitochondrial christae and the aggregation of ribosomes with amorphous materials suggests further reduction of metabolic coherence. The appearance of large vesicles suggests the fragmentation and reannealing of tonoplast and other membranes. Unfortunately, however, Meyer and Heath were unable to define ultrastructurally a diagnostic feature for cessation of cytoplasmic streaming.

The observations that browning of cells was a late indicator of death in cowpea cells and its failure to develop in tissue immersed in salt solutions bears further consideration. It is possible that the failure to develop brown discoloration reflects the fact, as noted by Meyer and Heath, that *all* of the treated cells were dying. Secondly, neighboring cells, are in this case inactive in phenyl propanoid synthesis, being most likely poisoned by heavy metals. Presumably the underlying healthy cells in fungus-infected tissue are the major source of such synthesis. Possibly, the process of cell death may have been too swift to permit browning to develop. However, Meyer and Heath suggest that HR induced in cowpea infected with *Erysiphe cichoracearum* is even faster than that observed in $CuCl_2$-treated cells; the microtubules and Golgi bodies disappeared faster and the membranes appeared more disorganized.

Apparently ultrastructural changes occur in cowpea that accurately predict the onset of cell death. However, an important dissimilarity between $CuCl_2$ and *E. cichoracearum* induction of cell death is the manner in which the fungus affects organellar membranes. Electron-opaque, lipid-like accretions appear in many membranes; mitochondrial membranes are thick and dark and mitochondrial christae are missing. Not only are these signs of membrane derangement, but they suggest a dissolution of the lipid bilayer and the coalescence of lipids into ultrastructural "pools". This is a common phenomenon in chloroplasts, in which the plastoglobuli increase in number and size as foliar tissues senesce (Goodman, Király and Wood, 1986) and has been seen in tobacco leaf chloroplasts in tissues undergoing HR induced by *Pseudomonas syringae* pv. *pisi* (Goodman 1972). Subsequently we shall present data revealing oxidation of membrane lipids by oxygen radicals that supports these observations.

Respiratory processes leading to melanization and gelation

The observations by Tomiyama (1955) of what appeared to be "gelation" or melanization of the cytoplasm requires additional consideration. Prior to the time that dead cells were observed, Tomiyama, Takakuwa, Takase and Saki (1959) detected in resistant potato tuber tissue infected with *P. infestans* an *increase in respiration 10-12% above that observed in susceptible tissue*. Inhibitor experiments suggested that the early increase in O_2 consumption was mediated by ascorbic acid oxidase and that *later*, when the invaded cells began to degenerate, a cyanide resistant respiration ensued. It is during the latter that stress metabolite synthesis of phytoalexins occurs in sliced potato tubers. It is, however, the earlier increase in O_2 consumption that we believe is critical to HR development. This reflects, according to Tomiyama and Ishizaka (1967), an increase in both the synthesis and oxidation of o-diphenols, ostensibly to their active quinoid form in which crosslinking of proteins in the cytoplasm may occur. The tissue browning, seen shortly after the inoculated cell is penetrated, reflects observed melanization and gelation which may occur through crosslinking proteins and polysaccharide polymers as suggested by Neukomm and Markwalder (1978).

Direct evidence for membrane damage

The HR-death of some potato tuber and leaf petiole cells occurs within 30 min of host cell wall penetration by *P. infestans* race 0 in cv Rishiri (Kitazawa, Inagaki and Tomiyama, 1973). The process of spore germination and prepenetration development of the hyphae requires about 50 min according to Sato *et al* (1971). Although most hyphae completed penetration of tuber cell walls in 130 min some were penetrated within minutes. Apparently the rate of penetration and subsequent symptom development depends on the metabolic state of the tissue. When freshly cut tubers were inoculated HR was apparent within 6 h, whereas, tissue that had been cut 20 h prior to inoculation (aged) demonstrated HR in 2 h. Apparently wounding affects some metabolic activities that potentiate HR development.

The rapidity with which host cell death developed suggested to Matsumoto, Tomiyama and Doke (1976) that death was a reflection of membrane damage. Their data in Fig. 5 reveal a continuously increasing flux of electrolytes from the time that race 0 penetrates the host cell wall. The loss of electrolyte from tissue in a compatible interaction also is continuous, but not as precipitous. However, when HR cell death was monitored microscopically by a plasmolysis procedure, it was apparent that the process developed at an exponential rate and that 90% of the infected cells died in 6 h. It seems that electrolyte leakage induced by a compatible race of *P. infestans* reflects membrane damage that is reparable whereas the damage caused by race 0 is irreversible.

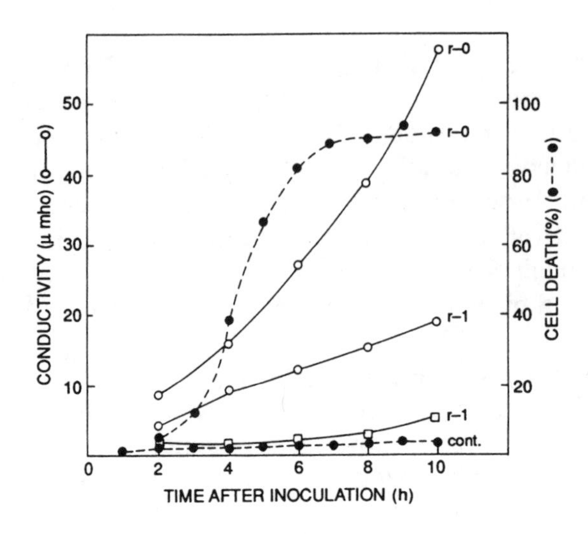

Fig. 5. Changes in electrical conductivity of water and cell death in tuber disks of Rishiri inoculated with incompatible race 0 and compatible race 1 of *P. infestans*. Inoculation was made 20 hr after cutting tuber disks. r - 0 = race 0, r - 1 = race 1. Modified from Matsumoto *et al.*, 1976.

A requirement for *de novo* protein synthesis?

Some studies from Professor Tomiyama's laboratory from 1957 to 1978 were directed at analyzing metabolic processes connected with HR that required *de novo* protein synthesis. It was possible to show that the uncoupling agents 2,4-dinitrophenol and azide applied to "aged" inoculated potato disks suppressed rapid cell death. However, it was possible to mitigate the suppression with ATP (but not ADP). Further, ATP treatment affected neither hyphal penetration nor intercellular hyphal growth, nor did ATP have any affect on the response of host cells in a compatible interaction. Hence, *it would appear that the uncoupling of ATP generation seriously affected HR development and that resupply of ATP supports its continued development.*

Particular attention is called however to the fact that in fresh non-aged disks inoculated with an HR-inducing race, ATP had only a slight effect on the time of hypersensitive death development. This suggests that during the period that tuber tissue disks age (5-16 h), some additional physiological factors develop that influence the HR potential of inoculated cells. Nozue, Tomiyama and Doke (1978) contend that only the metabolic activity that responds to ATP when uncoupling of phosphorylation is imposed, requires *de novo* protein synthesis. The nature of what might have been synthesized or altered is at this point a matter for conjecture. However, as we shall see there is good evidence that a primary site for HR-related disorientation is the plasma membrane and other membrane bound organelles. Perhaps the frequency or availability of receptor sites for elicitor molecules is involved.

The hypothesis that specific protein synthesis is required to support HR development was further tested by Doke and Tomiyama (1975) by applying the antibiotic blasticidin S (BcS). BcS is innocuous to *P. infestans* but suppresses protein synthesis of potato tissue by decreasing the activation of amino acids, the formation of aminoacyl-tRNA, and the incorporation of charged tRNA into protein (Gottlieb and Shaw, 1970). BcS at 10 μg/ml was applied to aged Rishiri leaf petioles which were then inoculated with race 0 of the pathogen. The antibiotic was not inhibitory to the fungus as penetration of host cells exceeded 90% in both BcS-treated and control tissue and there was no discernable effect of the antibiotic on the rate of infection. Protein synthesis was followed by measuring the incorporation of ^{3}H leucine into protein. The measurements recorded reduced ^{3}H leucine incorporation into protein in BcS-treated tissue as compared with the untreated control. In fact the ratio was approximately 9:1 for control vs. BcS-treated tissue. However, in tissue exposed to BcS for 0.5, 1.0 and 1.5 h before inoculation with race 0 zoospores, no diminution of HR was detected. Doke and Tomiyama interpreted these results as suggesting that *de novo* protein synthesis was not integral to the processes involved in HR death of potato cells infected by *P. infestans*.

In subsequent and related experiments Doke and Tomiyama (1977) examined the effect of SH-binding compounds on HR-death caused by race 0. The rationale for this study stemmed from the previously mentioned observations of Nozue, Tomiyama, and Doke (1978), that 2,4-DNP uncoupled phosphorylation in potato tubers and delayed HR. This delay, which was reversed by ATP (Fig. 6), showed that the synthesis of certain respiratory enzymes was essential to HR development.

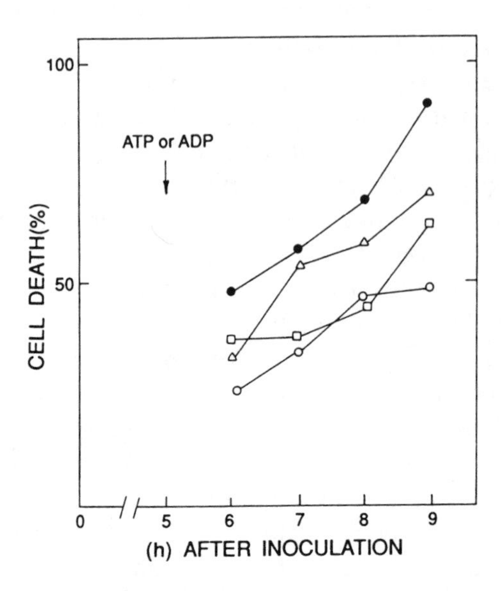

Fig. 6. Effect of 2.4-dinitrophenol (2.4-DNP), adenosine triphosphate (ATP), and adenosine diphosphate (ADP) on hypersensitive death of potato-tuber disks infected by the incompatible race 0 of *Phytophthora infestans.* The aged disks were treated with 0.1 mM 2.4-DNP for 30 min before inoculation. Inoculated disks were treated with 1 mM ATP (▲) or ADP (□) in 0.01 M phosphate buffer at pH 6.5 or the same buffer alone (O) 5 hr after inoculation. As a control, other disks were treated with water before inoculation and then treated with phosphate buffer (●). Nozue *et al.*, 1978.

The compounds, 2,4-DNP, DNFB, NEM and PCMP applied 30 min prior to inoculation altered neither the time of penetration nor hyphal growth of the fungus. However, they and PMDT (chloromercuribenzoic acid complexed with a high molecular weight dextran) all uniformly delayed the time required for HR-cell death to occur. In a second experiment the low molecular weight 2,4-DNP, DNFB and PCMP and the high molecular weight PMDT were applied to potato disks 6 h *after* inoculation. In this instance the browning observed 20 h after inoculation was inhibited by the low molecular weight reagents, but not by the high molecular weight PMDT. Doke and Tomiyama hypothesized that the large PMDT molecule could not penetrate the infected plant cell walls, enter the cell and influence browning (oxidation); nevertheless, it apparently delayed HR by affecting the plant cell membrane directly. Obviously the smaller SH-reagents entered the infected host cells, suppressed oxidative enzyme activity and in addition delayed the development of necrosis. It would seem that all of the SH-reagents influenced HR development by acting directly on the infected plant cell membrane. *These observations suggest that HR development per se (as indicated by plant cell death) is independent of the cell browning (necrosis) phenomenon that follows.* However, we are still without convincing data that suggest HR requires *de novo* protein synthesis.

Potentiation of the HR response of tissue by aging

As yet there is no convincing explanation for the potentiated HR activity in potato tuber or leaf cells that is gained by aging them for 5 - 16 h prior to inoculation with an HR-inducing race of *P. infestans.* Observations of this phenomenon were first made by Tomiyama (1960). His analysis at that time was that HR death usually takes place very quickly in metabolically active cells and more slowly in the inactive ones. Hence, *de novo* synthesis of a crucial protein appears to be important in the development of HR following the inoculation of potato tissue with *P. infestans* if HR reactivity is to be *potentiated.* This is accomplished either by inoculating potato tissue with an HR-inducing race of the pathogen and aging, as noted above, or by wounding the tissue and aging the exposed cells in a moist atmosphere at 18° C for 16 - 20 h. Clearly, the HR response is capable of being activated to a high level of potency as manifested by causing host cell death "minutes" after penetration of a resistant cultivar by pathogen with which it is incompatible.

Analogously, experiments by Furuichi *et al.* (1980b) presented evidence that even susceptible potato tissue, inoculated with race 1,2 which causes a compatible reaction displays a degree of HR-like resistance to infection if the inoculated tissue is young. Both petiole and tuber tissue, inoculated while in the juvenile state will express a degree of incompatibility. Similarly, Lisker and Kúc (1977) reported that both browning and terpenoid accumulation occurred in potato sprouts expressing compatible and incompatible interactions with races of *P. infestans.*

Number of potato cells contributing to the development of rapid HR

In order to examine the basis for potentiated HR capacity some comprehension of its source is necessary. We know that one measure of HR activity is the rate of potato cell death following inoculation with an HR-inducing strain of pathogen. Since penetration of host cell walls by isolates causing compatible or incompatible reactions is accomplished at the same rate, it is the subsequent development of the pathogen in the host cell that determines whether the host cell will die quickly or will support the continued growth of the pathogen.

In 1960 Tomiyama reported that an infected potato cell required the metabolic activity of an underlying 10 - 15 healthy cells to demonstrate a rapid HR. A subsequent publication by Kitazawa and Tomiyama (1973) established more precisely the number of underlying cells necessary to provide virtually complete inherent HR. The two studies were conducted with tuber tissue that had not been aged and therefore the data reflect inherent HR resistance. Infection was measured by the time taken for 50% of the cells infected in an incompatible interaction to die and the depth to which cell browning would develop in slices of varying thickness. Some pertinent data are given in Tab. 1.

Tab. 1. The relationship of potato tuber tissue thickness to HR-cell death and necrosis. (Compiled from three sets of data)

Tissue thickness in mm	Numbers of cell layers	Time for 50 % cell death (h)	Depth of brown lesion (μm)
0.8	5	32	800
1.5	11	22	184
2.5	18	-	168
5.0	35	16-17	145

(Penetration of all cells occurred in 130 min, disks were 2.2 cm in diameter, data recorded after 48 h) Kitazawa and Tomiyama, 1973.

It had been noted earlier (Tomiyama, 1960) that thinner slices of tuber tissue had lower resistance to infection. In addition, it had been observed in the earlier study that reduction of the spore load in the inoculum permitted very thin slices of tuber tissue to develop a high level of HR. Hence it would appear that the reduced disease in the thin slice was not caused by the slicing (wounding) but by the fewer numbers of underlying cells that were infected by the reduced spore load in the inoculum. More specifically, the data in Tab. 1 support the contention held by Tomiyama (1960) that the fewer the underlying healthy cells, the longer the time required for the development of hypersensitive death of the infected ones; and, the larger the resultant necrotic lesion. It would seem that the ideal number of cells underlying infected ones is around 18; however, this is a variable that could be significantly altered by reducing or increasing the spore load.

What then is the role that these underlying cells play? Some insight is provided by Kitazawa and Tomiyama (1973), who suggest that the physiological processes that inhibit the development of the disease lesion and also fungal development "occur after the death of the infected host cell". The observation that some cells die minutes after penetration would argue that HR is an expression of a gene or genes that requires little synthesis of metabolites by the resistant cell once recognition of the fungus has occurred. However, post-infection (host cell wall penetration), synthesis by the 18-35 underlying cells is required for lesion size and hyphal growth suppression.

The very interesting cinephotomicrographic observations made by Kitazawa, Inagaki and Tomiyama (1973) revealed that the cytoplasm in the infected cell accumulates more actively around the penetration sites in cells undergoing the incompatible reaction than those in cells exhibiting the compatible one. Accumulation of cytoplasm at the point of penetration by the fungus in the incompatible interaction suggests that *the potentiated metabolism, ie. accelerated and specifically directed cytoplasmic streaming and increased respiration are intimately associated with cell death.* Nishimura and Tomiyama (1978) studied membrane integrity of the tuber cells of R_1 Rishiri and susceptible Irish Cobbler at 1.4 and 2.4 hr after inoculation (prior to penetration and when 20% of the cells were being penetrated respectively) by testing cellular capacity to absorb ^{3}H leucine, $H_3{}^{32}PO_4$ and $^{86}RbCl$. At 2.4 h after inoculation the rates of ^{3}H-leucine and ^{32}P uptake decreased more rapidly in the cells infected by the incompatible pathogens than by the compatible one. At this stage no cell death had been recorded and hence it was concluded that the physiological activity of the plasmalemma and metabolism was seriously perturbed at a point in time coinciding with recognition by the resistant tuber cell of the HR-inducing fungal hyphae, *i.e.* during the early stages of penetration.

Do phytoalexins cause HR-induced host cell death?

A close examination of the singular phenomenon of the cessation of protoplasmic streaming is most revealing. It is reported to occur 10-20 min after fungal penetration in the incompatible interaction and not for at least 20 h after penetration in the compatible interaction. The time frame for cell wall penetration determined by Kitazawa, Inagaki and Tomiyama (1973) is 69-100 min after inoculation. As we shall see, the timing of these events stand in stark contrast with the timing of phytoalexin syntheses and accumulation.

Nevertheless, some authors have at least inferred that the HR may be a reflection of the accumulation of compounds such as rishitin. The relative efficacy of rishitin and related phytoalexin compounds such as lubimin and phytuberin, in inhibiting hyphal growth and influencing HR-related host cell death was studied by Sato and Tomiyama (1976a,b). Although some presumptions are made relative to the location of the phytoalexins in the tuber disk experiments, and the relative participation of phytoalexins other than rishitin, the computations appear reasonable. Two time course measurements are made in Fig. 7 indicating that 15 μg rishitin/g fw accumulates in the surface tissues of the cut leaf petioles of cv. Rishiri, inoculated with race 0 at 3 x 10^5 spores/ml, 15 h after inoculation. Initial detection of the phytoalexin was possible only after 7 h. At 15 h after inoculation it was estimated that rishitin had accumulated largely in the top third of the thin petiole slices (0.5 - 1.0 mm thick) and the amount of rishitin was calculated at 45 μg/g fw. A previous study by Ishiguri, Tomiyama, Katsui, Murai and Masamune (1968) found the ED_{50} for rishitin in suppressing hyphal growth of *P. infestans* to be 44 μg/ml.

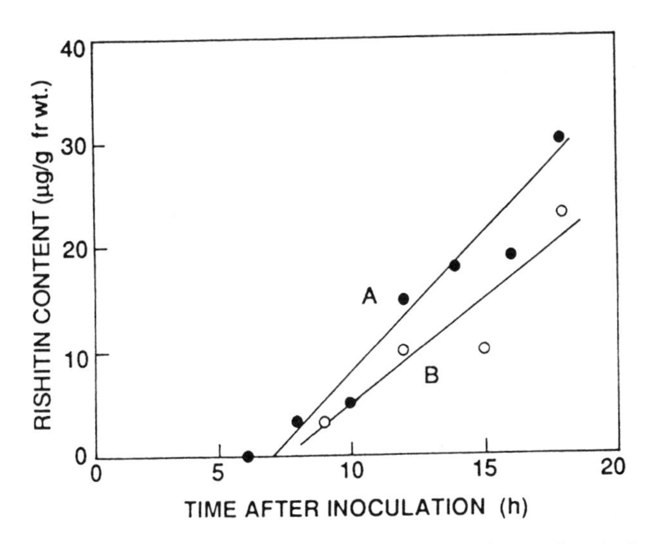

Fig. 7. Time course of rishitin accumulation in surface tissues of cut leaf petiole of potato cv. Rishiri infected by an incompatible race of *P. infestans* in two separate experiments, A and B. Sato and Tomiyama, 1976a.

Fig. 8 shows fungal hyphae of *P. infestans* in both compatible and incompatible interactions growing at parallel rates up to 7 h, when rishitin was first detected. However, point A on the curve indicates that by 3 h after inoculation almost all host cells inoculated with race 0 had died but none that were inoculated with race 1. It also is apparent from Fig. 8 that intercellular hyphal growth slows perceptibly as rishitin levels increase between 7 and 15 h after inoculation. However, the inhibition of disease lesion development in intercellular space and the suppression of hyphae invading neighboring cells required a level of rishitin of 100 μg/g fw according to Ishiguri *et al.* (1969). Hence, Sato and Tomiyama (I976b) concluded, quite logically, that rishitin plays an important role in suppressing fungal development in potato tissue. However, the accumulation of phytoalexins cannot be the cause of HR-cell death.

A similar analysis can be made from publications by Yoshikawa and Masago (1977) and Yoshikawa, Yamaguchi and Masago (1978), that in soybean infected by *Phytophthora megasperma* var. *sojae* (PMS), the phytoalexin glyceollin actively restricts the pathogen in resistant soybean hypocotyls. In the 1977 study, mRNA synthesis was followed with ^{3}H uredine pulse

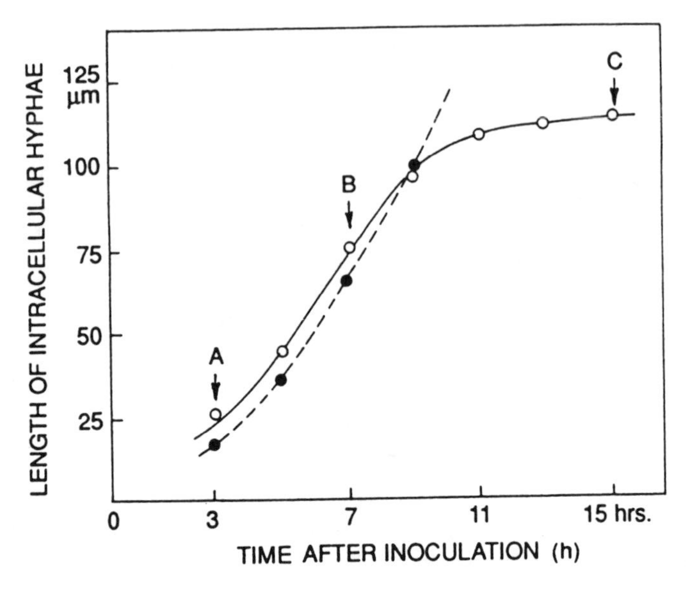

Fig. 8. Relationship between intracellular hyphal growth of *P. infestans* and rishitin content in an infected cell of cut surface tissue of potato leaf petiole. A: time when almost all of the host cells infected by race 0 have died, B: the time when rishitin began to be accumulated, C: the time when rishitin reached about the concentration of ED_{50} for hyphal growth inhibition *in vitro*. Tissue infected by incompatible (O) or compatible (●) race. Sato and Tomiyama, 1976b.

labelling in Harosoy 63 (resistant) soybean hypocotyls inoculated with races of PMS causing either incompatible or compatible interactions. Whereas both races stimulated mRNA synthesis, the rate was more rapid for the incompatible interaction. Whether these early-synthesized RNAs (4 h after inoculation and 10 h before resistance in the incompatible reaction) reflect ultimate rates of phytoalexin synthesis was not determined. The RNAs apparently influence resistance through the subsequent production of glyceollin which is produced in much greater amounts and more rapidly in cells exhibiting the incompatible interaction. In the subsequent paper of Yoshikawa *et al.* (I978) PMS grew at similar rates in Harasoy (susceptible) and Harasoy 63 (resistant) soybean hypocotyls for 9 h after inoculation, at which time glyceollin accumulated at localized infections at levels that exceed the ED_{90} for the phytoalexin. The authors concluded that glyceollin accumulation occurred in sufficiently high concentrations prior to the time that fungal growth ceased to account for the expression of cultivar-specific resistance of soybean hypocotyls to PMS.

Relationship of enzymes synthesizing phytoalexins to HR

The hypersensitive reaction is believed to be initiated by the interaction of an elicitor, of fungal origin, with the wall or plasma membrane of a plant cell (Bostock *et al.*, 1983). Recognition of the elicitor sets in motion a sequence of events that terminates with the rapid death of the plant cell. Associated with HR is the induction of increased synthesis of mRNAs coding for enzymes involved in subsequent phytoalexin synthesis. It has been shown however, that these enzymes, *e.g.* phenylalanine ammonia lyase, 4-coumarate:CoA lyase, chalcone synthase, catechol *O*-methyltransferase and others, are constitutive, and that elicitation causes their increase above base levels (Bell *et al.*, 1984, 1986). Nevertheless, it would appear that prior to elicitation, plants have a capacity for comprehensive aromatic synthesis and oxidative activity.

Role of haustorium in the HR

Meyer and Heath (1988b) also noted that *E. cichoracearum* inoculated into non-host cowpea appeared at first unsuccessful in cell penetration, inducing the development of a callose-like papillum. Subsequently the fungus developed a haustorium from a separate infection peg that killed the invaded cell. Apparently in this host-pathogen combination two resistance phenomena are induced: one the callose-like papillae that appear to restrict fungal entry and secondarily HR that is induced by the haustorial peg. The question is raised whether the initial unsuccessful attempt in some way lowered the cells' resistance, permitting infection to take place. It is also interesting to note that the successful penetration, nevertheless, caused a rapid HR resistance reaction.

Heath (1971) reported what appears to be a similar failed-infection by *Uromyces phaseoli* var. *vignae* in non-host *Phaseolus vulgaris.* The bean responded to the cowpea pathogen's initial attempt at entry by the deposition of an electron-opaque material, preventing haustorial formation in 90% of the infection sites.

The observations of rust infection on cowpea and bean emphasize the sage statement made by Heath (1976) that "rust resistance is obviously a complex phenomenon and the hypersensitive response, even if a primary determinant in restricting fungal growth, may only be the visible product of a number of previous interactions between the plant and its potential pathogen". Results of these "foiled" attempts at entry, however, were not detected in immune cowpea cultivars. In these, initial responses were quickly followed by the rapid and simultaneous disorganization of both haustorium and host cell. The sequence of events at least with cowpea rust depicts a much slower process of infection than was seen with *P. infestans* in potato. In host and non-host combinations haustoria were formed 12-24 h after inoculation and it is the living haustorium that induces HR 24-48 h after inoculation. In both resistant host and non-host cultivars, death of the haustorium and of the haustorial mother cell did not result in the immediate death of intercellular mycelium. Starvation or perhaps other resistance mechanisms completed the resistance reaction.

In her study of the effect of fungal death on the interaction between *Uromyces phaseoli* and resistant and susceptible bean cultivars, Heath, (1988), wrote that only the living haustorium elicited necrosis, doing so a few hours after it became mature. In these experiments the fungicide oxycarboxin killed the fungus, the intercellular hyphae and the haustoria; however polyoxin D, a chitin inhibitor, killed the intercellular hyphae but not the haustorium, because its walls do not contain chitin. The dead fungus, in either instance, elicited encasement of the haustorium in the absence of plant cell necrosis. By selectively inhibiting either the haustorium or intercellular hyphae it was possible to demonstrate that it was the *living haustorium* that caused necrosis. Additionally, the experiments revealed that killing the fungus with oxycarboxin within one day, but not two days after inoculation, prevented the development of necrosis that normally develops in a resistant cultivar. These results reinforce the findings of Heath (1984) discussed previously in this section. The observations also suggested that the haustorium was responsible for a uniform deposition of callose-like material along the walls of the invaded cell which may contribute to the death of the surrounding cells. It is in these adjacent cells that necrosis is detected first, about 48 h after inoculation.

Variations in HR development in host and non-host tissue

In monokaryon and dikaryon infections: In gene-for-gene systems involving biotrophic pathogens, active defense mechanisms are usually induced by avirulent but not virulent races (Chen and Heath, 1990). In these instances products of genes for avirulence (elicitors) react with products of resistance genes (receptors), resulting in the activation of a defense mechanism. Actually, no difference is seen in the initial response to fungal penetration between resistant and susceptible cowpea cultivars. However, HR induction varies perceptibly whether induced, for example in the case of the cowpea rust fungus, by the dikaryon of the fungus or its monokaryotic form. In the case of the former the interaction is activated by the formation of a haustorium in a subepidermal mesophyll cell. On the other hand, the first cell invaded by the monokaryon is commonly an epidermal cell and contact is with an intracellular primary hypha. Heath (1989) rightly calls attention to this latter infection process as one more easily monitored than subepidermal haustorial development. However, perhaps more important is the probable difference in response time by the attacked host cell. The proximity of the epidermal cell to the developing hypha suggests that the host cell has a temporal advantage in monokaryon elicited HR. Specifically, host cell death could come sooner than in the instance of subepidermal growth followed by the elaboration of a haustorium as is the case for the dikaryotic infection process. Heath (1989) suggests that the monokaryon makes its elicitors more rapidly available to the host cell. However, both mono- and dikaryon appear to activate the same genes for HR resistance as each of two resistant cultivars responded to the two infectious propagules in a characteristically similar manner.

In non-host tissue: The comparative nature of HR induced by biotrophic fungi in host and non-host tissue is not clear. The reactions may be disparate with respect to time, morphology and intensity. Hence the question is raised whether HR in these two instances are sufficiently distinct to reflect control by different genes.

In the case of the non-host Pinto bean inoculated with di-

and monokaryons of *Uromyces vignae,* the former generally failed to induce haustorium formation and in the latter instance most germ tubes failed to penetrate the epidermal cell (Heath, 1989). An earlier study by Heath (1972) comparing bean and cowpea infections by *U. vignae* suggested that haustorial formation in cowpea was implicit in signalling resistance (inducing HR) to dikaryotic infections. However, in Pinto bean, it seemed that the dikaryon-induced host cell wall modifications (electron opacity) prevented haustorial formation in all but 10% of the infection hyphae observed. Similarly, Leath and Rowell (1969) revealed that host cell wall changes such as mesophyll wall thickenings in corn inoculated with *Puccinia graminis*, appeared to prevent haustorial formation and hence precluded infection from developing. In fact some hyphae ceased growing before contacting the thickened mesophyll cell walls.

In a much later study Heath (1989) reported that Pinto bean infection with urediospores resulted in infection hyphae and haustorial mother cell development, but no haustoria were formed. When basidiospores were used as the inoculum, only appressoria and short primary hyphae developed. The plant response to the dikaryon was, generally, autofluorescence and increased refractivity of the mesophyll cell wall. The monokaryon inocula resulted in encasement of the primary hypha or death of the invaded cell. However, 60% of the germinated basidiospores failed to penetrate any epidermal cell. Clearly, the overall response of the non-host Pinto bean to either form of the cowpea rust was not typical of HR. It would seem that in these instances the bean plant has developed a mechanism that prevented the cowpea rust from forming the necessary infection structures and/or from detecting metabolites that are integral to HR elicitation. Apparently, growth of the fungus *per se* was somehow curtailed before it was able to induce HR.

It is unfortunate that the studies critically comparing dikaryon infections of host and non-host cells are few, and that time of induction of necrogenesis has not been precisely recorded. Until the gene product that triggers the HR in the above instances is elucidated, and a threshold level of it for activation of the phenomenon is determined, it seems that differentiating between the manifestations of HR in either host or non-host tissue will remain difficult if not impossible. However, it is clear that a more precise definition of HR can exclude the non-membrane-damaging features of resistance. What then remains is the possibility of characterizing HR as a spectrum of intensities that reflect the deciding biochemical elements that govern membrane disorientation. Thus until we know, for example, the precise gene(s) that control the level of superoxide generation, discriminating between host and non-host HR will remain an elusive matter. From recent reports, it would seem that the prevention of haustorium formation, prevention of germ tube penetration and phytoalexin synthesis are not features, *a priori,* of HR. Rather they are resistance mechanisms in their own rights, although phytoalexin synthesis may be linked and develops subsequent to HR.

HR reflects the sensitivity of mesophyll cells to *Erysiphe graminis* var. *hordei* haustoria

In a comparatively early paper examining HR in barley infected by *Erysiphe graminis* var. *hordei*, White and Baker (1954) contended that resistance to infection was dependent upon the speed of collapse in mesophyll cells underlying the haustorium-penetrated epidermal cells. Both mycelial development and the number of haustoria appeared to be limited by mesophyll collapse. All cultivars of barley: highly resistant, resistant, semiresistant and susceptible, supported haustorial development. Establishment of the pathogen was correlated with the number of haustoria that developed in epidermal cells. From each infection court in a highly resistant cultivar, one and rarely two haustoria formed in the epidermal cells under which 2-5 mesophyll cells collapsed as soon as a haustorium formed. In the resistant cultivar 5-12 haustoria formed in the epidermal cells beneath each colony, each causing small groups of mesophyll cells to collapse. The semiresistant cultivars developed 10-15 haustoria in the epidermal cells beneath each colony under which large groups of mesophyll cells collapsed. Abundant mycelium and conidia developed on the leaf surface of the susceptible cultivars. Here as many as 10-30 haustoria formed in the epidermal cells beneath each colony. However, the mesophyll cells appeared normal as no collapsed cells were detected. The number of haustoria that developed depended on the rapidity or degree of mesophyll collapse.

The time sequence in this host-pathogen interaction also is interesting. In the highly resistant cultivar, haustoria are formed in 30-48 h and all mycelial development is arrested within 54 h, when 6-8 mesophyll cells have collapsed. In the other cultivars haustoria formed initially within 54 h, and there was no collapse of mesophyll cells at that time. Hence, as compared to resistant cultivars, the response of mesophyll cells to the trauma, clearly induced by haustorium development, was markedly delayed in semiresistant cultivars.

It is clear that resistance of barley to *E. graminis* var *hordei* depends on the reaction of the mesophyll and not the epidermal cells. Thatcher (1942) has shown that permeability changes and respiration increases in mesophyll cells beneath the infection court. These physiological alterations are apparently related to the membrane damage that develops prior to mesophyll death and necrogenesis.

Host cell death precedes fungus suppression in *Bremia lactucae*-induced HR

Maclean, Sargent, Tommerup and Ingram (1974) observed that HR in plants infected by biotrophic fungi involves the rapid necrosis of a few cells at the site of penetration accompanied by the restriction of the fungus. It is likely that these are linked events; however, the primary determinant seems to be host cell death followed by necrotization of the invaded cells. Host cell death was indeed the primary event in their study of resistance in five lettuce cultivars to four races of the downy mildew pathogen *Bremia lactucae*. *Actually, death of the penetrated cell preceded the cessation of fungal growth by several hours.*

The time course of the infection process was followed by light microscopy and, as has been noted previously for both necrotrophic and biotrophic fungi, the germ tubes inducing both compatible and incompatible interactions, penetrated the epidermal cell walls in the same length of time. In this period of time both fungal races produced a primary vesicle and an elongated secondary vesicle. Whereas the primary vesicles developed at the same rate, the growth of the secondary had expanded to 5 x 9 μm in the incompatible reaction and was complete in 12 h. However, the secondary in the compatible interaction continued to expand for 24 h reaching a size of 14 x

26 μm. Over a time course of 60 h the compatible reaction caused no necrosis in any of the epidermal cells penetrated. However, the epidermal cells expressing the incompatible reaction developed progressive granulation and melanization. Fungal growth in the compatible interaction was characterized by extensive intercellular mycelial development bearing haustoria. In the incompatible reaction, the pathogen formed intercellular hyphae in 12 h, which stopped growing after 24 h following production of about three haustoria, each causing HR in the cells they penetrated.

Cytological examination of the hyphae of the two races revealed that the incompatible hyphal cell nuclei were unable to divide, indicating an inability to establish a biotrophic relationship. Perhaps rapid maturation of the intercellular hyphae and haustorial development account for the precocity of the HR-inducing race.

Electron microscopy of infected cotyledons expressing a compatible interaction revealed invagination of the host cell plasmalemma by both primary and secondary vesicles without evidence of cytoplasmic damage. After 4 h, however, penetration in the instance of the incompatible interaction caused disorientation of membranes and cytoplasm of epidermal cells. Fungal cytoplasm within the incompatible interaction was, visually at least, normal at this time.

In summary, HR is apparent in resistant cultivars in 24 h, whereas, susceptible cultivars reveal no visual damage in 60 h. The observable damage is a disruption of host cell membranes and cytoplasm in response to an apparently normal invasive haustorium.

HR is a point response of the plasmalemma to haustoria of *Puccinia graminis*

Invaded host cells of a resistant line of wheat containing the *Sr6* gene for *P. graminis tritici* resistance developed clear signs of a susceptible reaction when the infection developed at 26° C (Harder *et al.*, 1979). Moving the infected seedling to 19° C did not preclude cells at the periphery of the rust lesions from becoming infected. However, these newly infected cells expressed an incompatible reaction and were highly autofluorescent (evidence of necrosis). Cells that were clearly infected at 26° C and were then exposed to 19° C failed to develop signs of incompatibility. *The authors interpreted this to mean that the significant events determining incompatibility occur during the development of the haustorium and the invagination of the host plasmalemma.*

According to Harder *et al.*, (1979), the primary indicator of HR in infected leaf cells was disruption of the host plasmalemma, particularly at the region of invagination. *Hence a major feature of HR is a "point response" to the haustorium.* Ultrastructural variations develop at sites of direct contact between the invaginated host plasmalemma and the intruding haustorium and cause HR. These variations include distant *vs.* close association of host and haustorial membranes, interrupted *vs.* continuous host cell plasmalemma and vacuolated *vs.* normal cytoplasm of infected mesophyll cells.

Incompatibility, as manifested in the pathogen, is initially expressed as a uniform increase in cytosol density. This increase is apparent first in haustoria and then in the haustorial mother cells. Again, the effect appears to occur sooner at the site of contact.

The effect of temperature on HR development in Puccinia-infected wheat

Although the study by Harder *et al.* (1979) emphasizes certain ultrastructural features of HR as it develops in wheat infected by *P. graminis tritici,* their data establish the temperature sensitivity of the genes involved in this resistance reaction. When mesophyll cells are invaded by an HR-inducing strain at 26° C, no autofluorescence (necrosis) develops. However, when the experiment is conducted at 19° C, or plants inoculated and infected at 26° C are subsequently transferred to 19° C, the cells not previously infected become necrotic upon invasion. Similar results were reported earlier by Silverman (1959) also on wheat stem rust and by Zimmer and Schafer (1961) for *Puccinia coronata* on oats. Hence, HR is severely limited when infection proceeds at elevated temperatures; in fact the resistant reaction noted at 19° C becomes at 26° C a classical susceptible reaction. This control of HR by temperature has been reported in plants inoculated with either viral or bacterial pathogens capable of causing incompatible reactions. Although some of the genes involved are known, their specific gene products remain unknown.

Elicitor and receptor of the HR phenomenon

Two important features of HR that still require discussion are: a) the nature of the substance or substances that trigger or elicit rapid cell death and necrosis and b) the target site and composition of the receptor for the eliciting action.

It had been known for some time that cell free hyphal components of *P. infestans,* extracted from either virulent or avirulent races induce necrosis in both susceptible and resistant cultivars (Tomiyama, 1971; Varns *et al.*, 1971). However, Doke (1975) presented convincing evidence that HR, induced in potato tuber or petiolar tissue of three resistant cultivars by race 0 of *P. infestans,* was prevented or delayed by prior treatment with a high molecular weight fraction obtained from zoospore homogenates, Fig. 9. Although it was apparent that fungal wall components

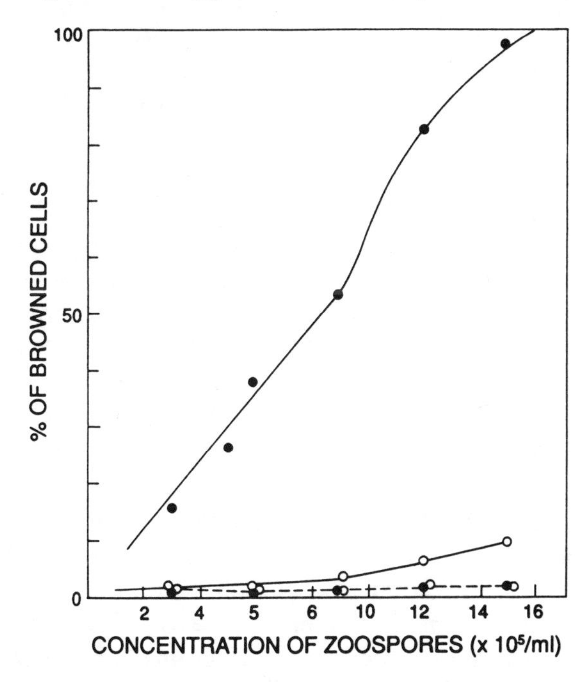

Fig. 9. Effects of zoospore concentration in the inoculum and as a source of extracts on cellular browning. Tissue disks, 18mm in diameter, were inoculated by 50 μl of zoospore suspension at the concentration shown or treated with 50 μl of extract derived from a zoospore suspension of similar concentration. About 1000 cells, in total, from two different experiments were observed under a microscope 24 h after inoculation or treatment. (—) Inoculated; (---) treated; (●) race 0; (O) race 1. Doke, 1975.

from either virulent or avirulent races induced HR, this study raised the question concerning possibile specificity between zoospore homogenates and proteinaceous substances from susceptible (Irish Cobbler) and resistant (Rishiri) potatoes.

Binding of fungal components to host cell

The importance of pathogen hyphal binding to the host cell protoplast was visualized with light microscopy by Nozue, Tomiyama and Doke (1979). Potato tuber cells inoculated with either race 0 or race 1 became infected and were subsequently plasmolyzed in a hypertonic sucrose solution. A tent-type plasmolysis was observed; i.e. the wall-penetrating hyphae (of either race) adhered so tightly to the plasmolyzing protoplast that at point of contact, the retraction of the protoplast was restrained. As a consequence the plasmolyzed protoplast appeared tent shaped. Where the protoplast retracted, fragments adhered to the infection hypha. Little difference was detected in the cohesion of plasmalemma to either race 0 or race 1 hyphae. It was clear however, that adherence between the two surfaces was comprehensive from the earliest moments of infection (*i.e.* contact).

Nozue *et al.* (1980b,c) presented evidence that potato lectin might play a role in binding the infecting hypha to host membrane. This report received support from evidence developed by Furuichi *et al.* (1980a) and more recently from Furuichi and Suzuki (1990) that lectin from potato binds to hyphal wall components (HWC) of *P. infestans.* In the latter study no differences were observed between compatible and incompatible combinations. However, a specific inhibitor of lectin-binding inhibited HR caused by an incompatible race.

Influence of hyphal wall components (HWC) on potato protoplasts

Doke and Tomiyama (1980a, b) examined the sensitivity of protoplasts from a number of potato cultivars containing an array of resistance genes for *P. infestans.* Seven races of the pathogen were tested, utilizing the technique with which Tomiyama, Lee and Doke (1974) had found protoplasts to be sensitive to HWC. The 1980 experiments revealed that there were no significant differences in the physiological activities of the were no significant differences in the physiological activities of the wall components of the 7 races on the protoplasts. Each caused rapid protoplasmic aggregation, as had been noticed in the earliest of observations of cyclosis (Tomiyama, 1955), and lysis of the protoplast.

There were, however, significant differences among the protoplasts with respect to their reactivity to both the source and concentration of HWC. Specifically, the higher the *field resistance* of the cultivar, *the greater the reactivity* of the protoplast to HWC. Maximum protoplast reactivity was noted at 125 μg HWC/ml. However, as little as 1 μg/ml was sufficient to initiate reactivity of protoplasts to HWC. The maximum rate of reactivity to HWC was within 5-10 min of time of application. It seemed therefore that *the recognition or target sites that respond to the HWC (elicitor) causing HR are on the plasma membrane.* From this study it would appear that protoplasts of a cultivar exhibit a rate of reactivity and a degree of sensitivity to HWC that positively correlates with its resistance to *P. infestans*, at least in the case of field resistance.

The correlation of HR to resistance *per se* is given added support by the experiments of Doke and Furuichi (1982) who monitored the response of protoplasts from 12 potato cultivars to HWC from HR-inducing race 0. The percentage of reactive protoplasts from each cultivar was closely correlated ($r = 0.942$) with the percentage of dead cells caused by infection of tuber slices with race 0 (Fig. 10). Specifically the cultivars having the smaller number of cells invaded by race 0 had the greater percentage of protoplasts reacting with HWC. The correlation between HR protoplasts responding to HWC and the number of dead and browned tissue cells inoculated with *P. infestans*, an inverse relationship, was $r=0.912$ ($P< 0.001$) (Fig. 11).

These data suggest that HR is not an all or nothing response, but probably requires a threshold level of gene product for expression. Additionally, it would appear that the incompatible recognition reaction of host cells with some HWC triggers a sequence of processes that culminate in the development of HR resistance in either protoplasts or tissues of potato.

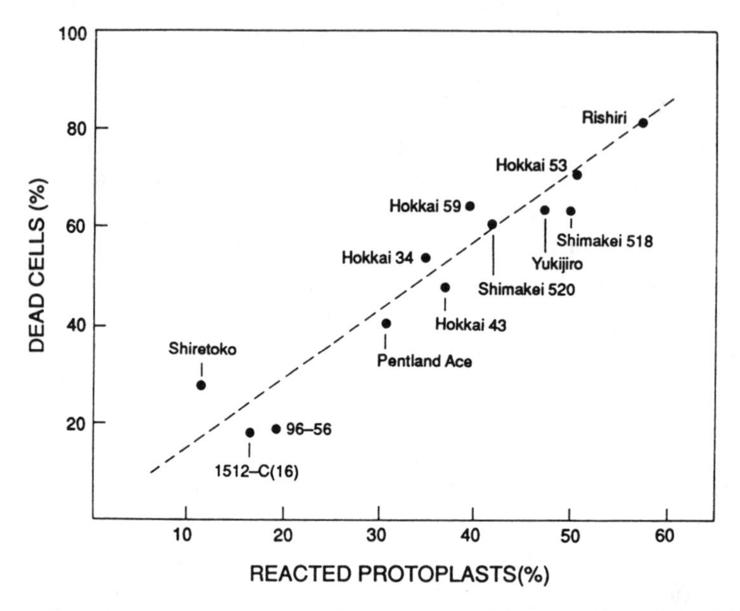

Fig. 10. Correlation between the percentage of hypersensitive cell death in potato tuber tissues infected by an incompatible race (race 0) of *Phytophthora infestans* and the percentage of hypersensitively reacted protoplasts of potato tuber tissues to hyphal wall components of the fungus. r=0.942 (P<0·001). The cell death in the first infected cells was observed 7 h after inoculation of aged tuber discs (17 mm in diameter, 2 mm thick) with 100 μl of zoospore suspension at a concentration of 100 x 10^4 ml^{-1}. The protoplasts from each cultivar were exposed to hyphal wall component at a concentration of 0.5 mg ml^{-1} for more than 30 min. Each value for the percentage of reacted protoplasts and dead cells in tissues was derived from observations of about 600 protoplasts and 300 infected cells in 2 experiments, respectively. Doke and Furuichi, 1982.

The hypersensitivity inhibiting factor (HIF)

An improved understanding of the chemical nature of the compound that inhibits the HR (HIF) was developed by Doke, Garas and Kuć (1979). In experiments in which HR was induced in the resistant cv. Kennebec, HIF was extracted from *P. infestans* race 1.2.3.4. These "factors" were characterized as **β** 1,3 and **β** 1,6 glucans with polymeric lengths of 17-23 glucose units. The glucans, derived from both mycelia and zoospores, were non-anionic and anionic polymers with the latter having one or two residues with a phosphoryl monoester.

The HR inhibitor suppressed both death of host cells and browning as well as the accumulation of rishitin in tuber slices inoculated with race 4 of the pathogen. The glucans from the races causing compatible reactions were more active in suppressing HR than those from races capable of causing incompatible reactions and anionic glucans were more efficacious than were non-anionic ones.

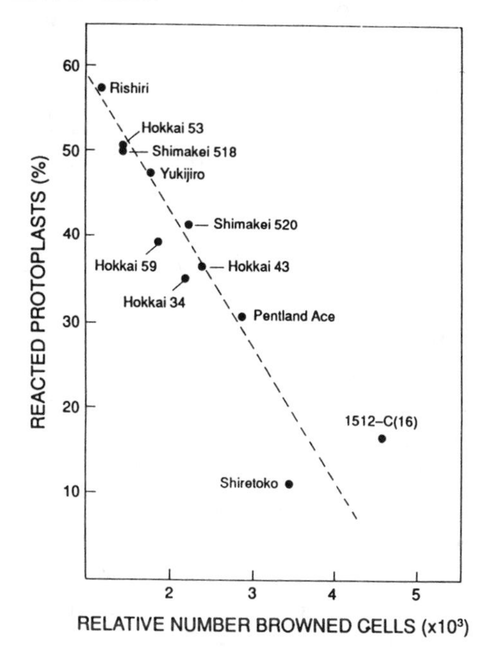

Fig. 11. Correlation between the percentage of hypersensitively reacted protoplasts of potato tuber tissues to hyphal wall components of *Phytophthora infestans* and the relative number of dead and browned cells in inoculated potato tuber tissue used as a resistance index r=0.912 (P<0·001). Doke and Furuichi, 1982.

Crude elicitor (HR-inducing) preparations from races 4 and 1.2.3.4 lost their terpenoid (phytoalexin) inducing activity when mixed with microsomal fractions of potato tuber tissue. Of particular importance was the discovery that glucans from the race causing a compatible reaction, but not the one causing an incompatible one, reduced this loss of activity. It appeared, therefore, that the *compatible reaction* between potato tissue and *P. infestans might result from the suppression of HR* by water soluble glucans produced by races of the fungus that cause a susceptible reaction.

In one experiment, 90% of the control cells in potato tuber tissue inoculated with race 0 of *P. infestans* died within 18 h, whereas 30 and 70% of the infected cells pretreated with glucans from race 1.2.3.4 (compatible) and race 4 (incompatible), respectively, died. In a related experiment, protoplasts isolated from Rishiri were mixed with glucans from race 1 (compatible) and race 4 (incompatible) for 15 min, washed to remove excess glucan, and then exposed to hyphal wall components. Of the protoplasts treated with race 1 glucans 45.5% were protected against HR, whereas only 3.5% of those treated with race 4 glucan survived. It is essential to emphasize the HIF suppressed not only HR cell death, but also reduced electrolyte leakage, cell browning (necrosis) and the synthesis of terpenoids like rishitin in infected cells.

In summary it would seem that HIF suppresses the reaction between fungal elicitor and host cell plasmalemma. Essentially, an elicitor from zoospores or hyphal wall tissue of *P. infestans* is blocked from reaching a host plasmalemma receptor by a polysaccharide produced by races of the pathogen that cause a compatible reaction.

The efficacy of the pathogen in suppressing HR may derive at least partially from the fact that HIF is released at the time of spore germination, and at an appropriate moment to prevent HR induction. With respect to HIF being released by *P. infestans* zoospores, mention should be made concerning the study by Doke and Tomiyama (1977) wherein a procedure was developed to synchronize encystment and germination of zoospores. Shaking the $CaCl_2$-containing spore suspension optimized the process, engendered encystment and germination in 2 h, and facilitated the collection of glucans from the germination fluid.

A more recent study by Furuichi and Suzuki (1990), however, showed that there was little difference in effect on HR or HR-suppression by glucans extracted from races capable of inducing either compatible and incompatible reactions. Nevertheless, it remains certain that HIF inhibits HR.

Elicitation of HR by arachidonic and eicosapentaenoic acid

The active components in the crude elicitor, HWC, described by the Tomiyama group were chemically identified by

Bostock, Kuć and Laine (1981). Precise purification procedures indicated that eicosapentaenoic acid (EPA) and arachidonic acid (AA) are the highly active components of the crude elicitor preparations from *P. infestans* mycelial and spore suspensions. In fact, Maniara *et al.* (1984) detected elicitor activity at the picomole level when accompanied by glucan enhancers. Comparisons with a wide array of eighteen saturated and unsaturated fatty acids suggested EPA and AA to be the resistance-triggering entities. Removal of the fatty acids from HWC by methanolysis, and hexane extraction resulted in loss of eliciting activity (Bostock *et al.* 1981).

Attention is called once again to the procedure for inducing HR to *P. infestans* in potato tuber slices. An essential feature of the process is the wounding necessary to obtain cylindrical slices. As a consequence of wounding, the tissue is stimulated to synthesize and accumulate steroid glycoalkaloids (SGA). However, it was observed by Tjamos and Kuć (1982) that treating potato slices with AA or EPA inhibited the accumulation of SGAs and instead prompted the accumulation of the phytoalexin sesquiterpenoid compounds rishitin, lubimin (and others). *Hence in addition to inducing the HR characteristics such as leakage of electrolytes, rapid but restricted host cell death, browning, and the subsequent accumulation of phytoalexins, AA inhibited SGA synthesis and accumulation.* Thus the elicitors had actually altered important host metabolic pathways, concerned with specific aromatic ring compound synthesis. Tjamos and Kuć (1982) also reported that the application of AA to slices (at 100 μg/ml) almost completely inhibited sterol synthesis even after SGA synthesis had been in progress for 120 h.

It is of additional interest that in the experiments done by the Kuć group, potato tuber tissue was aged (6 h at 20° C) prior to initiating their elicitor experiments. As previously discussed pretreatment in this fashion is apparently necessary to amplify metabolic events that condition the HR response. This pretreatment also was found essential in studies reported by the Tomiyama group (Furuichi *et al.* 1979a). Similarly, Bostock *et al.* (1983) noted that slices from tubers stored for 1 month at 4° C and inoculated with an HR-inducing race of *P. infestans* generated higher levels of phytoalexin, their measure of HR intensity, than potatoes stored at 25° C.

Potentiation of AA and EPA elicitor activity by glucans (Bostock, Laine and Kuć, 1982) of specific conformation adds another dimension to the induction of HR. According to Preisig and Kuć, 1985, the inability of glucans *per se* to induce any of the cascade of events set in motion by either AA or EPA, suggests their *modus operandi* to be one of regulating the initiation of the response. Whether this concept involves activation of the acyl forms or mobilizing, in some fashion, the acyl form to an enzyme site remains unresolved.

The candidate enzyme for freeing AA or EPA from its in-wall acyl form is lipoxygenase which is plentiful in potato tissue. Preisig and Kuć (1987a) have reported that salicylhydroxamic acid (SHAM), which inhibits the activity of lipoxygenase, prevents both AA and EPA from inducing HR in potato tissue. They suggest that the specific activity of lipoxygenase on the substrates AA and EPA in acyl form, and the suppression of the enzyme by SHAM, link the enzyme to the induction of HR by AA and EPA.

In a study by Bostock *et al.* (1993) lipoxygenase (LOX) activity was increased 2-fold in potato tuber disks by AA; the increase was detected within 30 min, peaked in 1-3 h, and returned to control levels 6 h after treatment. The activity is predominantly from a 105kD protein, which conforms in size to other lipoxygenases. When the enzyme was reacted with potato tuber extracts, 5-hydroperoxy-eicosatetraenoic acid was the principal product.

LOX activity was not potentiated by non-elicitor fatty acids or by β-glucans of *P. infestans* mycelia. However, pretreatment of the disks with ABA, SHAM, and N-propyl gallate, all of which suppress AA induction of HR, also inhibited the AA-induced facet of LOX activity. Importantly, cycloheximide which abolishes other AA elicited responses in potato tissue, does not inhibit LOX activity in these tissues.

From the above it is possible to consider that 5-LOX activity is a response to AA that is coupled to HR. Bostock *et al.* are uncertain of the direct causal relationship between lipoxygenase activity and HR because they failed to detect a concomitant increase in LOX activity in these studies. However, recent data from the Bostock laboratory (personal communication) revealed a greater than 3- fold increase in LOX in potato leaves inoculated with *P. infestans* sporangia (Fig. 12).

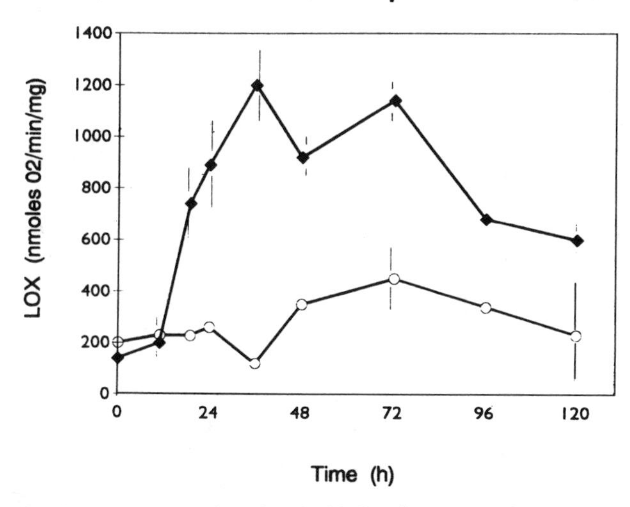

Fig. 12. The potato leaves inoculated with *P. infestans* revealed a greater than 3-fold increase in 5-LOX activity in 24 h. Bostock, personal communication, 1993.

Interaction between elicitor (AA), lipoxygenase and peroxidase in HR development

A more precise indication of the linking of AA to the induction of HR in potato through lipoxygenase has been developed by Vaughn and Lulai (1992). Two potato tuber callus lines, one lacking lipoxygenase activity (LOX^-) and a second possessing activity (LOX^+), were treated with AA at 20 μg/ml. The LOX^+ calli exhibited some HR death and browning of the outer layers of cells, those that had come in contact with AA, within one h of treatment. However, the LOX^- calli did not exhibit changes until 24 h had elapsed (Fig. 13). By that time the LOX^+ calli exhibited 60% cell death and browning.

Of particular importance was the steep increase in peroxidase activity of more than twice control levels demonstrated by the LOX^+ calli six h after treatment with AA (Fig. 14). During this same period the LOX^- calli showed a decrease in peroxidase activity. The authors suggested that peroxidase may be necessary for the development of the browning reaction that occurs as a consequence of HR. Peroxidase may also be associated with the early increase in oxygen consumption in

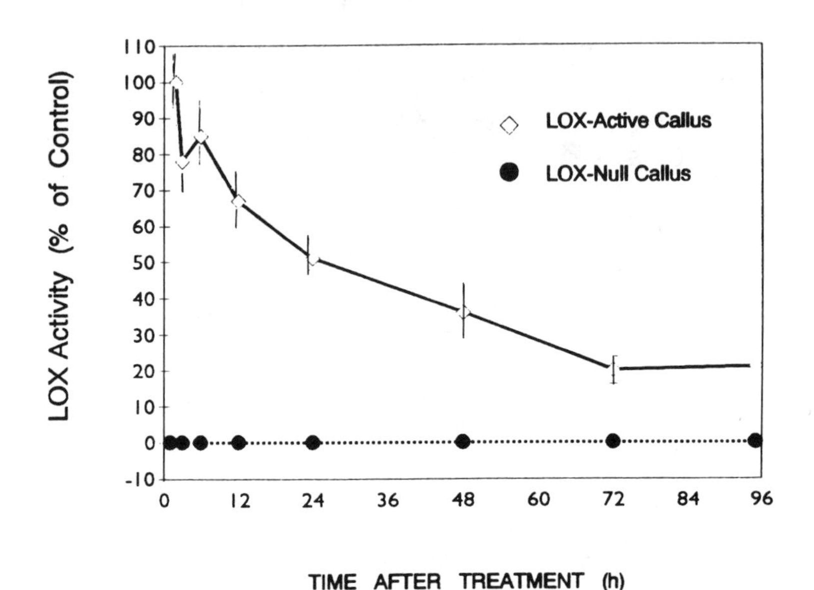

Fig. 13. LOX activity in two potato tuber callus lines, LOX^+ and LOX^-, that were treated with 20 μg/ml AA to induce HR. Vaughn and Lulai, 1992.

H_2O_2-related membrane damage. Since tissue browning occurred only in LOX^+ calli, it would appear that this feature of HR is due to the increased peroxidase following stimulation of LOX activity. This could, ostensibly, occur as the result of superoxide synthesis.

Interesting also were their data that suggested that polyphenol oxidase (PPO) seemed not to be correlated with the observed browning caused by AA. Even though the LOX^- calli did not brown they developed more than three times the PPO activity in response to AA treatment than control tissues. Hence, although AA potentiated PPO activity in calli from both tissue culture lines, this enzyme appeared not to be associated with typical HR-related symptoms. Attention of the reader is called to repeated historical reference of PPO to induced plant resistance reactions to disease. However, it is also to be recalled that PPO-related resistance to disease has been reported to be induced by an array of elicitors and has its basis in a number of metabolic pathways (Goodman, Király and Wood, 1986).

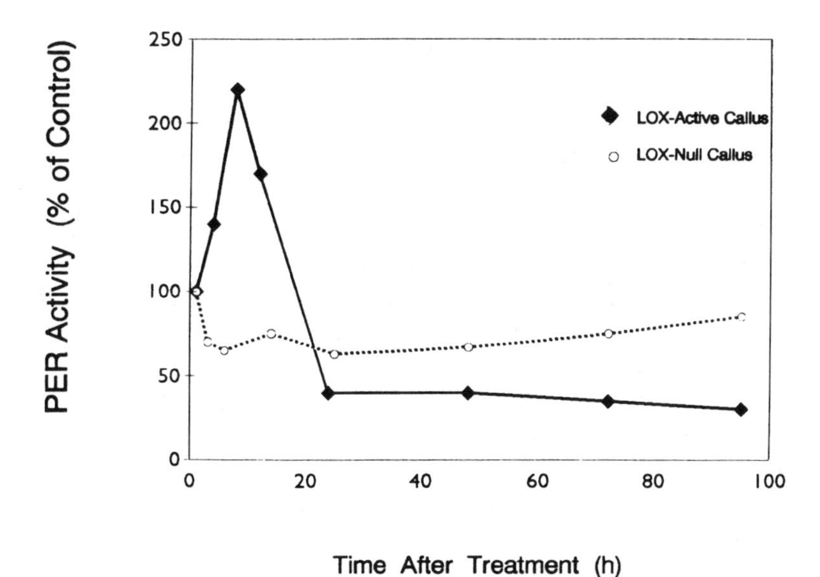

Fig. 14. Peroxidase activity in two lines of potato tuber callus, Lox^+ and LOX^-, that were treated with 20 μg/ml AA to induce HR. Vaughn and Lulai, 1992.

The relationship of lipoxygenase to light in controlling HR-related necrosis

In experiments by Peever and Higgins (1989) electrolyte leakage, lipoxygenase activity, and lipid peroxidation were induced by both specific (SE) and non-specific (NSE) elicitors from *Cladosporium fulvum* in resistant but not susceptible cultivars of tomato. It was also noted that lipoxygenase activity was induced as early as six hours after injecting tomato tissue with either elicitor. The SE was prepared from intercellular fluids of tomato leaves infected with race 2.4.5 of *C. fulvum*, whereas the NSE was prepared from culture filtrates of the pathogen. It was of additional interest that although both elicitors induced electrolyte leakage, their differential response to lipoxygenase inhibitors suggested that they were chemically unrelated elicitor compounds with similar biological activity. The specificity of SE was further demonstrated as it induced lipoxygenase activity in the resistant cultivar Sonatine, but not in the susceptible Bonny Best.

Peever and Higgins made a critical observation regarding the necrosis feature of the HR syndrome induced by SE in tomato, and not by NSE: *the injected tissue required at least a brief light exposure for necrosis development. This observation suggested that membrane alteration and the necrosis facets of HR are separable in this interaction.* It may also be inferred from these data that products induced by SE are photosensitized in the light which results in the transfer of energy (electrons) to oxygen or other molecules ($O_2^{\cdot-}$, H_2O_2, phenols *etc.)* which are in fact necessary for the development of necrosis. A similar sequence of events was reported by de Wit *et al.* in 1988, also in a resistant tomato cultivar inoculated with *C. fulvum.*

The essentiality of necrosis in the development of HR-resistance has also been questioned in HR induced by bacteria. Electron micrographs revealed (Goodman, 1972) that membrane damage induced in tobacco by the incompatible *P. s.* pv. *pisi* developed in the absence of necrosis as long as the injected tissue was kept highly hydrated.

In a subsequent report by Vera-Estrella *et al.* 1992 using the Peever and Higgins system, several lines of evidence were presented that suggested the involvement of $O_2^{\cdot-}$, H_2O_2 and $OH^{\cdot}$ in the incompatible or HR resistance reaction. They further concluded that the production of active oxygen compounds is extracellular. The early appearance of peroxidases in the cells elicited an efflux of intercellular fluids from infected tissues which suggested an active role for these enzymes in the process.

Glucans, enhancers and suppressors of HR

It would appear from the study by Maniara *et al.* (1984) that the HIF glucans that supress HR as reported by Garas *et al.* (1979) and Doke *et al.* (1979b) and the enhancer glucans of Preisig and Kuć (1985) are one and the same. They are, in both instances, glucose polymers, probably β1→3 and 1→6 linked, and are branched chains of 17-23 units in length. Preisig and Kuć (1987b) suggest that the enhancement of HR is a non-specific phenomenon, whereas suppressor capability is race specific with respect to R gene resistance in potato infected by *P. infestans*.

The basis for their dual activity is not fully understood (Maniara *et al.*, 1984; Preisig and Kuć, 1987b); however the differences may reflect polymer length, degree of branching and concentration differences. The recent report by Furuichi and Suzuki (1990) defining the properties of purified glucan suppressor revealed the activity of two moieties with MW of 4700 and 280. Both were detected in races 0 and 1.2.3.4 of *P. infestans*, that induce incompatible and compatible reactions, respectively, in cultivar Rishiri. The data were inconclusive whether enhancement and suppression were caused by the same compounds and it was suggested that their concentration might influence specificity.

Still another possible explanation involves the bound or free state of the fatty acid elicitors. For example, the suppressor glucans may block a binding site in the host cell membrane which is the receptor for carbohydrate-bound forms of AA and EPA. Analogously, the enhancer could increase the activity of lipoxygenases which might catalyze the formation of more active forms of free AA and EPA. It is known for example that the free form of AA elicits HR more rapidly than the esterified or bound form of the fatty acid. This would suggest that release of AA from its native esterified form in the fungus is a necessary first step in eliciting HR. Caution is warranted, however, as Doke has shown that although AA stimulates sesquiterpene accumulation, it *does not* trigger superoxide anion generation (Doke *et al.,* 1987). The latter, as we shall see subsequently, is a candidate for the cause of membrane damage associated with HR.

Elicitor activity of AA is enhanced not only by glucans, but by Ca^{2+} as well (Zook *et al.*, 1987). It has become apparent that Ca^{2+} plays a role in the transduction of a number of stimuli that impinge upon the earliest stages of HR development. The manner in which this cation acts as a second messenger is not at all clear. As we shall see, Ca^{2+} appears to control the production of H_2O_2 which is also an early player in three critical features of HR: oxygen radical generation, membrane damage and electrolyte leakage.

The apparent dual function of abscisic acid on induced resistance

Just as glucans have appeared to both potentiate and suppress the overt symptoms of HR, so too does abscisic acid (ABA). Henfling *et al.* (1980) reported that ABA inhibited elicitor-induced phytoalexin accumulation in potato slices causing the treated tissue to exhibit a susceptible rather than a resistant response. We emphasize here that the effect was the suppression of the development of browning symptoms and a reduction in terpene accumulation. Subsequently Bostock *et al.* (1983)

confirmed this report and noted that the efficacy of ABA in suppressing sesquiterpene production depended upon low temperature 5-month storage (aging) of the tissue prior to treatment with ABA. A study by Cahill and Ward (1989) noted that decreases in ABA in soybean inoculated with *P. megasperma* f. sp. *glycinea* (PMG) appear causally associated with incompatible reactions, browning and desiccation of affected tissue. The ABA levels were monitored by ELISA and are of interest because ABA decreases are detected so early (after 2 h) following inoculation with PMG. It is between 2-4 h that membrane damage associated with HR in PMG-infected soybean tissue develops. In companion experiments a decline in ABA in tissue reflecting a susceptible reaction occurred only after 24 h and was due to fungal colonization of tissue.

A study by Ladyzhenskaya *et al.* (1991) suggests that necrosis *associated* with HR induced by arachodonic acid (AA) is fully suppressed by ABA. The mode of action is ascribed to the blocking of AA from its site of action, a membrane bound ATPase (apparently the proton pump ATPase). What is perplexing is that the influence of AA can be negated by ABA applied 6-8 h after AA treatment. Recognizing that the induction of and early developments of HR in potato tissue elicited by either *P. infestans* or AA is accomplished in 2-4 h, we conclude that ABA influences the development of necrosis and subsequent terpene synthesis (Henfling *et al.*, 1980 and Bostock *et al.*, 1983), but not HR *per se*. Less clear is the reason for the early decreases in ABA noted by Cahill and Ward (1989). They offer explanations based on loss of ABA by leakage through disrupted membranes and transport away from the infection site.

In contrast, the experience of Dunn *et al.* (1990) with induced resistance against *Colletotrichum lindemuthianum* in *Phaseolus vulgaris* suggested that necrosis was linked to *increases* in ABA. The ABA inhibitor fluridone decreased the concentration of the hormone 60% and increased symptom severity. The linkage between increased ABA levels and potentiated PAL activity was offered as a possible *raison d'etre*.

We call the readers attention to the foregoing opposite affects of ABA and suggest that they may reflect differences in concentration of this hormone and the overall hormonal balance at the site of action. There are numerous reports of the dual action of hormones wherein low concentrations stimulate and higher ones suppress a specific plant response. However, we also remind the reader that accentuated necrosis and the subsequent accumulation of products synthesized via shikimic and mevalonic acid pathways are features of plant metabolism that *follow* the induction and early phases of HR development.

A hypothetical sequence of events explaining the observation of Dunn *et al.* (1990) would be that a primary stimulus by ABA causes Ca^{2+} to be released from binding sites in the wall or "free space". Ca^{2+} then enters the cytosol and binds directly to a response element, for example, calmodulin (a protein modulator). This non-protein complex then binds to the response element (receptor) for example peroxidase. The magnitude of the response is, according to Owen (1988), proportional to the concentration of the Ca^{2+} - calmodulin - response element. Another possible scenario is that Ca^{2+}, stimulated by ABA interacts at the plasmalemma opening specific Ca^{2+} channels causing the influx of Ca^{2+} into the cytosol. Other schemes envision potentiated phosphorylation of protein(s) at the membrane that modulates their sensitivity to Ca^{2+} and potentiates entry of that ion into the cytosol (Schwacke and Hager, 1992)

Membrane potential changes induced by HR

Again we must turn to the case history for *Phytophthora infestans,* and the perception of Kohei and Tomiyama for the initial studies on membrane potential changes caused by HR-inducing fungi.

In their experiments, Tomiyama, Okamoto and Katou, (1983), studied the effect of race 0 of *P. infestans* and race 1, 2 which cause incompatible and compatible reactions, respectively, in potato petiole and tuber cells of cv. Rishiri. In the conventional procedure, a single microelectrode positioned with a micromanipulator is used to impale a plant cell while under microscopic observation. A second or reference microelectrode outside the cell in the tissue bathing solution measures changes in membrane electropotential, E_m, as a consequence of inoculation and the subsequent passage of time. The technique differentiates the two components of E_m. These are E_d, the diffusion potential, that part of the membrane potential that accounts for free non-energy requiring ion movement through the membrane and E_p or electrogenic (pump) potential, that component which is respiratory energy-requiring and under the control of a membrane-embedded ATPase. These experiments, conducted under anoxic conditions, could be performed under rather prolonged periods of surveillance.

It was observed that as fungal penetration occurred no change was evoked in E_m by the incompatible race; however, shortly thereafter E_d began to decrease. The E_p increased, (hyperpolarized) slightly compensating for the decrease in E_d which resulted in a slow decrease in E_m. Subsequently, E_p and E_m depolarized gradually, but from one to several hours after penetration the impaled cell depolarized rapidly. This was evident *before the cell contents granulated*. The E_d and E_p of the uninfected cell did not alter during more than 10 h of surveillance. Similarly, the cells infected by the fungal race which caused a compatible reaction revealed no membrane change during the 24 h course of the experiment. From the foregoing it is clear that the incompatible race of *P. infestans* decreased E_d soon after or just before penetration. Light microscopic examination of the sequence of events could not pinpoint precisely the time of penetration. Attention should be called to the fact that the early depolarization of the membrane had two effects, first increasing the activity of the *energy dependent* pump potential and second, an increasing membrane permeability. Both of these suggest alterations to membrane structure and will be discussed more fully in the "Bacteria" section of this monograph.

The tissues used in these experiments were aged for 20 h prior to cell impalement and inoculation, hence the tissues had generated their full "hypersensitivity potential". Nevertheless, neither race evoked an instantaneous membrane response. In earlier experiments ^{86}Rb efflux and influx was accelerated when about 20% of appressoria had begun to penetrate the cell wall (Nishimura and Tomiyama, 1978b). The depolarization of diffusion potential that was detected therefore may reflect permeability changes in the membrane caused by contact between hyphal and plasmalemma surfaces. The increase in permeability may actually signal K^+ fluxes of the type which were detected by Matsumoto, Tomiyama and Doke (1976) and Atkinson *et al.* (1989, 1990).

More recent electropotential measurements by Tomiyama and Okamoto (1989) were directed at determining whether depolarization of the E_d of a penetrated cell would occur if HR cell death was in some way inhibited. Previous studies had shown that the damage to cells resulting in HR (and cell death) appears to occur just before or immediately after the infection hyphae enter the host cell. Spore germination and movement of the hyphae through the cell wall have little effect in either compatible and incompatible combinations; *P. infestans* in both instances requires the same length of time (69-100 min) to traverse the cell wall thickness. Hence, the crucial sequence of events that triggers and fosters the development of HR occurs once the hypha of the pathogen contacts the host cell plasmalemma *per se*. Tomiyama and Okamoto (1989) reasoned that if depolarization of E_d always accompanies HR-cell death, then treatments which prevent or delay HR would preclude the development of membrane depolarization.

Pre-inoculation heat treatment of tuber tissue delays HR significantly (Tomiyama, 1957a). Heat treatment at 47° C for 5 min had little effect on E_d of control potato cells; but it protected cells penetrated by the HR-inducing race from any changes in E_d. Heat treatment also delayed and sometimes prevented tissue browning.

Thus the most distinct and general electrophysiological response involved in HR may be the depolarization of E_d, the passive diffusion component of E_m. In incompatible interactions, this appears to occur almost immediately after recognition at the plasma membrane surface, although the physiological processes of HR require additional time for cell death to occur.

Oxidation inhibitors such as 2,4-DNP, NaN_3 and several sulfhydryl compounds also delayed or suppressed HR (Doke and Tomiyama, 1977; Tomiyama, 1957b). However, these affect the energy dependent component E_p of the membrane. These observations tend to support the report of Nozue *et al.* (1978).

Although no direct evidence for the participation of the superoxide radical in the depolarization of E_d was shown (Tomiyama and Okamoto, 1989), the participation of malonate-sensitive respiration was inferred. Specifically, as glycolysis and the TCA cycle become increasingly inoperative, the substrate for oxidation is lipid rather than carbohydrate. Tomiyama and

Okamoto used the combined data of Romberger and Norton (1961) and Furuichi *et al.* (1979a) to present an apparent relationship between host cell death and increases in malonate-sensitive respiration (Fig. 15).

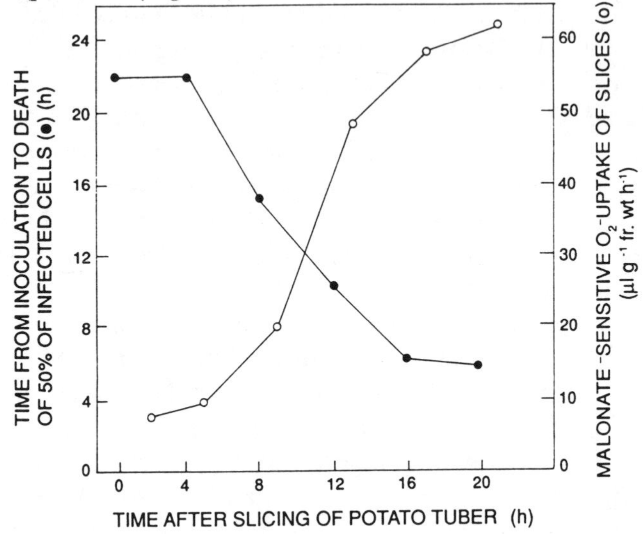

Fig. 15. Relation between malonate-sensitive O_2 uptake and the period of time from cell penetration by the incompatible race of *P. infestans* to hypersensitive death of the penetrated cell in potato tuber slices aged for different periods. The O_2-uptake cited is from Romberger and Norton (1961) and the time from penetration to hypersensitive death of 50% of the penetrated cells is cited from Furuichi *et al.* (1979a). Tomiyama and Okamoto, 1989.

The role of oxygen radicals in HR induction

Although our concept of how oxygen radicals and their generation play a role in HR will be defined in the "Summary" chapter, some basic facts are presented here. For a more comprehensive essay on the subject of free radicals the reader is directed to Leshem (1988). Sufficient background will be presented here to assist the reader in understanding research that has recently been published on oxygen radicals and their relationship to HR.

The earliest report on a possible role for oxygen free radicals in plant physiology, linked them to senescence and came from the laboratory of Dr. Barry Commoner, Washington University. The phytotoxic free radicals that have been recognized include $O_2^{\cdot-}$, $\cdot OH^-$, the peroxyl $ROO^\cdot$, polyunsaturated fatty acids (PUFA) and semiquinone free radicals.

Leshem (1988) revealed that the $O_2^{\cdot-}$ is generated both by insult to plant tissue and in normal cellular oxidative processes. In the presence of superoxide dismutase (SOD), the $O_2^{\cdot-}$ is converted to H_2O_2. In the presence of catalase or glutathione reductase, the H_2O_2 is converted to O_2 and H_2O. Both $O_2^{\cdot-}$ and H_2O_2 can be toxic to tissues; however, singly their toxicity is nullified because the aforementioned $O_2^{\cdot-}$ scavengers and peroxidative enzymes are prevelent. As noted by Taylor and Townsley (1986), when $O_2^{\cdot-}$ and H_2O_2 combine in what is known as the Haber-Weiss reaction, the exceedingly reactive $^{\cdot}OH$ radical is generated.

$$1) \quad \text{tissue damage} \longrightarrow O_2^{\cdot-} \longrightarrow H_2O_2$$

$$2) \quad O_2^{\cdot-} + H_2O_2 \longrightarrow \underset{\text{Reaction}}{\text{Haber-Weiss}} \longrightarrow {}^{\cdot}OH$$

The schematic shown above offers some experimental rationale for testing the possible toxicity of $OH^{\cdot}$ radicals, for example by attempting to mediate their effect with either mannitol, or preferably with the in-tissue very mobile DMSO. Clearly the impact of $O_2^{\cdot-}$ radicals can be abated by the infusion of SOD into plant tissue suspected of being damaged by the superoxide. Other blocking agents that might be evaluated in this respect are catalase or peroxidases.

Doke (1983a) first presented evidence that the superoxide anion was directly involved in the generation of HR in potato tuber tissue infected by *P. infestans*. Both nitroblue tetrazolium (NBT) and cytochrome *c* (cyt *c*), which make it possible to visualize nanomolar quantities of $O_2^{\cdot-}$ (Taylor and Townsley 1986), were actively reduced by potato tuber tissue exhibiting resistant, but not a susceptible interaction. In fact, Doke detected active NBT reduction around invasion sites 15 min after fungal penetration. Although this study did not precisely locate the site of $O_2^{\cdot-}$ generation, NBT, which is able to permeate cells, was observed in the reduced formazan state at the point of contact between the invading hyphae and the plasmalemma prior to host cell death.

Oxygen radical activity also was demonstrated in tissues exposed to HWC (See Fig. 16). Additional evidence for the participation of oxygen radicals in inducing HR was derived from data showing that SOD inhibited the reduction of either NBT or

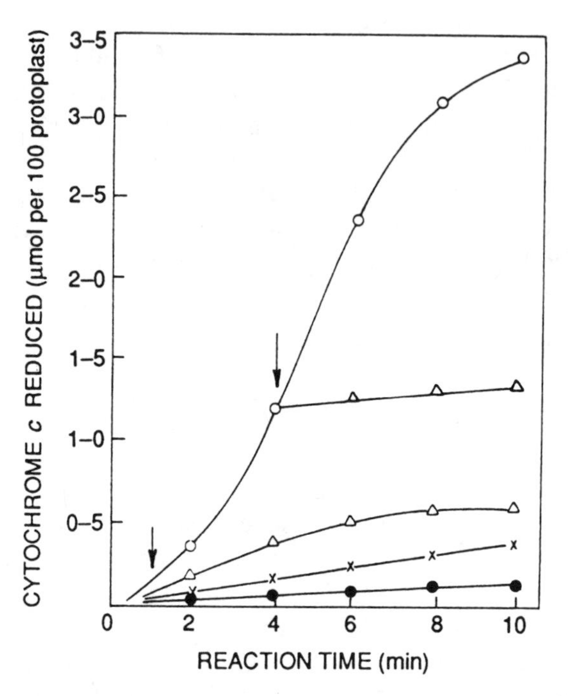

Fig. 16. Reduction of extracellular cytochrome *c* by protoplasts from potato tuber tissues following treatment with hyphal wall components (HWC) of *Phytophthora infestans* and inhibition of the reduction by superoxide dismutase (SOD). Complete reaction mixture consisted of 0.001M phosphate buffer, pH 7.8, 0.6M mannitol, 0.1 M EDTA/ 10 µM NADPH, 10 µM cytochrome *c* and protoplasts (500 to 600/ml). The reaction was initiated by adding HWC isolated from race 0 of *P. infestans* (125 µg/ml, final concentration). SOD (100 µg/ml, final concentration was included in the reaction mixture. O, Complete reaction system minus SOD; Δ, SOD was added to complete reaction system at the time shown by arrows; x, complete reaction system without NADPH; ●, no elicitation by HWC in complete reaction system. Doke, 1983b.

cyt *c* in the assay solution. In addition, HR cell death could be delayed when inoculated tissues were infiltrated with SOD.

Doke also determined that inhibitors of rapid HR death and depressed phytoalexin accumulation, such as SH reagents and NADP (Doke 1985; Doke 1983b), concomitantly inhibited $O_2^{\cdot-}$ generation. It was concluded that $O_2^{\cdot-}$ generation was involved in HR death induction; however, the precise form of active oxygen radical remains unclear.

NADPH oxidase, a basis for an elevated HR potential?

The almost axiomatic maximum HR potential of aged disks as contrasted with the lower potential in fresh potato disks tends to support the contention that *maximum HR potential is a reflection of an elevated capacity to generate active oxygen radicals.* As had been predicted earlier (Tomiyama, Doke and Lee, 1971), HR potential is enhanced by the generation of respiratory energy through the acts of wounding and/or infection with *P. infestans* resulting in resistant but not susceptible interactions. Doke (1985) established that NADPH-dependent cyt *c* reducing activity in a membrane-rich fraction of potato tuber tissues was enhanced by a) wounding (wound induced activity) (WIA), b) by infection with *P. infestans* races that induce resistant but not susceptible interactions (infection induced activity) (IIA) (Fig. 17) and c) by treatment with HWC.

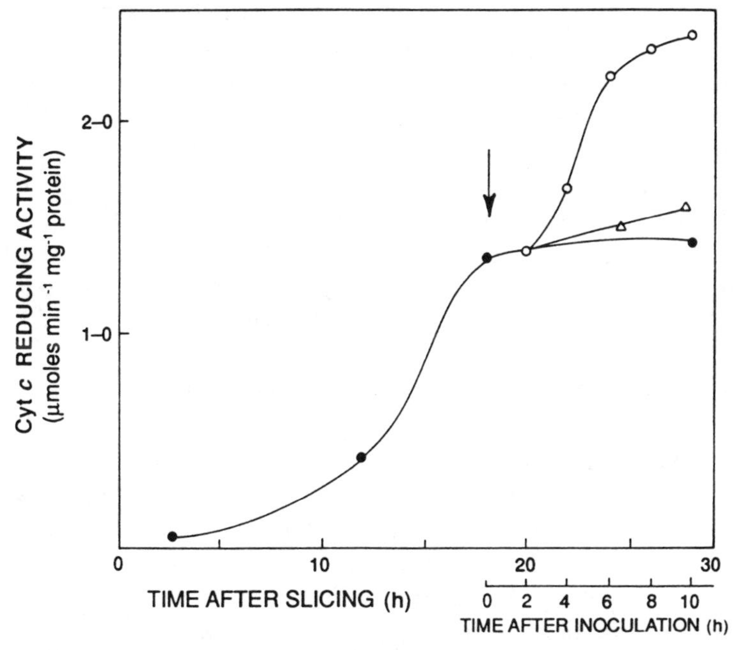

Fig. 17. Time course changes in NADPH-dependent cytochrome *c* (native) reducing activity in a membrane-rich fraction isolated from potato tuber tissues (cv. Rishiri, R_1) following wounding or inoculation by compatible or incompatible races of *Phytophthora infestans*. Tissues were incubated 18 h after slicing and inoculation with a zoospore suspension of race 0 or race 1.2.3.4 at a concentration of 50 x 10^5/ml 18 h after slicing (arrow). (●), Uninoculated; (O), inoculated with race 0 (incompatible); (Δ), inoculated with race 1.2.3.4 (compatible). Doke, 1985.

IIA was significantly inhibited by SOD and $NADP^+$, which suggested that an NADPH-dependant $O_2^{\cdot-}$ generating system may be activated in the membrane fraction from potato tissue undergoing the resistant reaction. Similarly, membrane fractions from tissue that had been treated with HWC or inoculated with an HR-inducing race also possessed a cyt *c* reducing activity that was in turn sensitive to SOD. Tissue exhibiting a susceptible reaction revealed no elevated cyt *c* reducing activity.

Although WIA reflected an NADPH-dependency, its response to acetylated cyt *c* and SOD suggested that it was activated by an NADPH-dependent cyt *c* reductase other than the one that activates IIA in this regard. Cyt *c* P450 was hypothesized by Rich & Lamb (1977). It is also possible that the WIA reflects the recovery of Krebs cycle and glycolysis types of respiration following utilization of lipid substrate. Perhaps, during the wounding process, the tissue switches substrate from carbohydrate to lipid (Day *et al.*, 1980) because lipids are released during wounding from the cell membrane, and the substrate activates lipid acyl hydrolases and lipoxygenase. As this substrate is utilized, the tissue returns to normal respiratory pathways, which concomitantly provide for a maximum HR potential. Whether these are causally related or only coincidental events remains unknown. Hasson and Laties (1976) suggest that it is the free fatty acids released by lipid hydrolysis in slice-injured potatoes that inhibit elements of the tricarboxylic acid cycle.

The role of NADPH oxidase in HR development

The report by Doke (1985) indicates that plants contain an $O_2^{\cdot-}$ generating NADPH oxidase. The time course during which the NADPH oxidase system appears to reach maximum activity in isolated membrane fractions is 8 h after exposure of the tissue to the pathogen, when 100% of the penetrated cells have died. However, $O_2^{\cdot-}$ generation in tissues is most active at 3-4 h after inoculation. Hence, *it would appear that the earlier burst of* $O_2^{\cdot-}$ *generation (between 2 and 4 h after inoculation, Doke, 1985, Fig. 17) accounts for cell death, whereas the continued oxidative activity has an additional agenda,* perhaps more closely related to phytoalexin synthesis by adjacent healthy cells. Attention is also called to the increase in NADPH oxidase activity that develops 2 h after slicing injury and continues to increase for about 18 h.

This is clearly a non-specific increase and reflects the WIA defined by Doke in Fig. 17. *We suggest that this increase and its related synthesis of active oxygen species may be related to the non-specific "oxygen burst" we will describe subsequently (Low et al., 1986).*

The secondary stimulus for the activation of NADPH oxidase appears to be the contact between penetrating hyphae and host cell plasmalemma. This assumption is based on the activation of the enzyme approximately 2 h after inoculation of potato tuber tissue. *Thus activation of NADPH oxidase may be one of the earliest, if not the initial, biological response of the HR phenomenon.* This assumption seems justifiable since germination and movement through the host cell wall by the infection hyphae of either resistant or susceptible interaction proceed at the same rate and without recognizable influence on HR development.

Oxygen radical generation in other hosts

Doke (1985) mentioned that in preliminary results with soybean, bean, pea, tomato, pepper and other plants, an incompatibility appeared capable of activating an $O_2^{\cdot -}$ generating system. He conjectured that superoxide activity is not limited to lipid oxidation of membranes, but may also play a role in reducing substrates such as polyphenols, phytoalexins and lignin precursors (*e.g.* cinnamic acid derivatives). In preliminary experiments it was found that inhibitors of $O_2^{\cdot -}$ generating oxidases such as phospholipase A_2, fatty acid cyclooxygenase and lipoxygenase also suppressed phytoalexin synthesis in HR-reacting potato tuber tissue. Hence it is not surprising that $O_2^{\cdot -}$ generation in potato tuber tissue is linked to AA and EPA as elicitors of phytoalexin in *P. infestans* infection (Bostock *et al.*, 1981, 1982).

In this regard Murai (1987) proposed that the fatty acids EPA and AA do not elicit HR directly (monitored as phytoalexin production). Rather, H_2O_2 is generated by the hydroperoxides that are produced when the fatty acids react with lipoxygenase. Murai suggested that the H_2O_2 might be derived from triplet oxygen directly or from superoxide. Since H_2O_2 is the most stable of the active oxygen species, it could act as a common trigger of HR as will be discussed in a subsequent section.

Role of water-soluble glucans (WSG)

Of importance is the previously discussed revelation that water soluble glucans (WSG) derived from hyphae and zoospores of *P. infestans* races causing *susceptible* reactions readily suppress HR. It seems possible then, that the susceptible reaction includes elaboration of a glucan that either turns off HR or precludes the triggering or recognition feature that fosters activation of NADPH oxidase. Some additional insight into this question may be derived from experiments in which hyphal wall components (HWC) and WSG were applied to protoplasts or to membrane fractions from tuber tissue exhibiting a resistant reaction (Doke, 1983b and 1985). When WSG was permitted to react with protoplasts *prior* to adding the HWC to the reaction mixture, NADPH oxidation of cytochrome *c* was almost totally suppressed. If, however, the WSG reacted with the protoplasts 5 min *after* HWC was administered, then the suppressive effect of the glucan was greatly diminished. On the other hand, if a membrane fraction derived from resistant cv. Rishiri, raised to full HR potential, was 7 h into the infection process when it was exposed to WSG, NADPH oxidation of cyt *c* was again almost totally prevented. *It would appear that this late continued activation of NADPH oxidation of cyt c is unrelated to HR and is perhaps, more closely associated with phytoalexin synthesis.*

These effects may reflect differences in glucan polymer length or influence on different enzyme receptors (cyt *c* dependent NADPH oxidases). Doke (1985) suggests that there may be separate sites for WSG and HWC and that the WSG site remained exposed on the membrane fragment, but was hidden or cryptic on the protoplast membrane after the enzyme (NADPH oxidase) was activated by HWC. However, when the WSG is added to the reaction mixture 15 min prior to the addition of NADPH, the glucan may have prevented NADPH from reaching the oxidase enzyme site. Similarly, when WSG was able to reach the protoplast membrane 5 min before HWC, activation of NADPH oxidase and HR were both suppressed. Hence, the glucan may simply act in both instances as a site-blocking agent.

Activation of NADPH oxidase

Studying the activation of NADPH oxidase biochemically is certainly a route to a better understanding of both the elicitors and the receptors involved in HR. In addition, studying potentiated hypersensitivity and induced resistance to infection following oxidase activation are logical research objectives. In this respect Doke and Chai (1985) evaluated compounds that would meet these objectives and found digitonin, a steroid glycoalkaloid, to be a competent activator of NADPH oxidase. Solutions containing 0.5% ethanol with digitonin at 100 μg/ml were applied to upper and lower leaf surfaces, upper surfaces of tuber disks, petioles and protoplast suspensions.

All tissues and protoplasts responded to digitonin by activating an extracellular cyt *c* reducing system. Evidence of the $O_2^{\cdot-}$ generating system was shown and this activity could be greatly depressed by the oxygen radical scavenging enzyme superoxide dismutase (SOD). Additional evidence for the activity of NADPH oxidase was its dependence on a supply of the reduced nucleotide in the protoplast system and its inhibition by oxidized $NADP^+$. Superoxide generation also was inhibited in the potato protoplast system by *p*-chloromercuribenzoic acid and N-ethylmaleimide. The absolute dependence on an exogenous supply of NADPH by protoplasts, as opposed to potato leaf and tuber tissue, suggested to Doke and Chai (1985) a minimal pentose phosphate pathway activity in protoplasts. It is indeed logical to assume that pentose pathway activity, which is generally stimulated under conditions of stress or aging, is a potent participant in the plant cells' oxidative metabolism. The pentose pathway supports the shikimic acid pathway and the synthesis of phenylalanine, tyrosine and their derivatives (Goodman, Király and Wood, 1986). It should also be kept in mind that potentiation of the pentose pathway is indeed a non-specific response to both trauma and senescence in plants.

Digitonin treatments also protected potato leaves from infection by *P. infestans.* This effect was presumed due to the influence of the elevated levels of $O_2^{\cdot-}$ on the infecting fungus. In fact, *a steep rise in $O_2^{\cdot-}$ was detected in both potato tuber and leaf tissue in an hour and within 2 min in protoplast suspension cultures.* In effect, digitonin elevated the HR potential in treated tissues by generating $O_2^{\cdot-}$ rapidly. When tuber disks were treated with digitonin at 100 μg/ml, no antifungal compounds were found in the tissues nor was the phytoalexin rishitin detected. However, in the susceptible reactions the fungus did elicit phytoalexin accumulation in digitonin-treated disks, but not in untreated ones. These results tend to suggest that early tissue responses in resistant reactions, that elicit HR, *are distinct from those that foster phytoalexin synthesis.* This was the conclusion that Ishizaka *et al.*, (1969) and Sato and Tomiyama (1976b) reached earlier (see page 24).

The induction of superoxide anion generation in potato leaves was again examined by Chai and Doke (1987), who found the process to develop in two stages. The first was *the transient activation of* $O_2^{\cdot-}$ *generation by fungal spores or their germination fluids prior to fungus penetration.* This is race-cultivar nonspecific and was induced by *P. infestans* races in both compatible and incompatible combinations. The *second stage which appears to vigorously reestablish itself at about 4 h after inoculation* continues for at least an additional 6 h with increasing intensity in reducing activity. *This increase was exhibited exclusively by tissues developing a resistant interaction.* The $O_2^{\cdot-}$ generation was closely associated with the occurrence of hypersensitive cell death. These experiments provide the possibility of discriminating between very early non-specific "oxygen burst" leading to $O_2^{\cdot-}$ or H_2O_2 generation and a subsequent specific elaboration of active oxygen radicals in a resistant reaction.

Evidence for the hydroxyl radical involvement in HR-related phytoalexin accumulation

Epperlein *et al.* (1986) used $AgNO_3$ at 10 mM to induce phytoalexin accumulation in soybean cotyledons and attempted to mimic glyceollin synthesis with superoxide generated by xanthine oxidase. Although appreciable levels of glyceollin were elicited by $AgNO_3$, $O_2^{\cdot-}$ generated by the enzyme was unable to elicit accumulation of phytoalexin. In addition, the glyceollin production elicited by $AgNO_3$ was not suppressed by SOD. They surmised that glyceollin was not synthesized under the influence of $O_2^{\cdot-}$. However, in related experiments, the $OH^{\cdot}$ scavengers DMSO, benzoate, mannitol and methionine all inhibited glyceollin production. Benzoate also inhibited pisatin accumulation in pea cotyledons that had been incubated with $AgNO_3$. These results suggested that $OH^{\cdot}$ is a mediator in abiotic elicitation of phytoalexins in soybean and perhaps in pea, and that $O_2^{\cdot-}$ is not directly involved in glyceollin production.

In this and other experiments we have some hesitancy in interpreting data concerning the development of HR when phytoalexin accumulation is the sole indicator of the process. It would have been more convincing if the data of Epperlein *et al.* (1986) had included observations that included alterations in host cell membrane integrity.

Some doubt in the oxygen radical-HR hypothesis

The data presented by Doke (1983a, b, 1985), Doke and Chai (1985) and Chai and Doke (1987) have provided compelling evidence for the involvement of superoxide anion in the development of HR in potato tissues and protoplasts by *P. infestans* and its hyphal wall components. The concept has received added support from the report by Doke and Ohashi of an $O_2^{\cdot-}$ -generating system in HR induced by TMV in tobacco. The theme has also been given credence by Keppler and Novacky (1987) who have accredited superoxide with HR-related membrane damage in cucumber, caused by *Pseudomonas pisi*. A convincing role for oxygen radical generation in HR development in response to cucumber mosaic virus infection was reported by Kato and Misawa in 1976.

However, as might be expected, the course of science rarely runs unrestrictedly as Moreau and Osman (1989) were unable to reproduce the potato - *P. infestans* experiments of Doke. Departures from the procedures used by Doke included tuber tissue that was not aged and the cultivar Kennebec rather than Rishiri. Perhaps the most significant variation in the Moreau and Osman study is their use of non-aged tissue. As has been noted, Tomiyama and his associates emphasize the importance of age-potentiated tissues for HR induction. Apparently aging amplifies the HR response several fold.

Jordan and DeVay (1990) studied plasma membrane and lysosome disruption in cells of Kennebec potato leaves inoculated with *P. infestans*. Their ultrahistochemical assessment of membrane damage was based on detecting electron-dense reaction products of acid phosphatase which were more prevalent in resistant than susceptible infections. Although their leaf disks were aged for 18 h before inoculation, the generation of superoxide and hydroxyl free radicals was greater in susceptible than in HR reacting tissues. However, there is a vast difference between peripherally wounded leaf disks and the exposed injured surface of tuber disks used by Doke and Chai (1985). The tuber disk experiments also showed superoxide initially in both susceptible and resistant reactions. Of interest in the Jordan and DeVay study was the observation that SOD suppressed superoxide but increased hydroxyl free radical accumulation in both susceptible and resistant interactions. However, the level of $OH^{\cdot}$ radical detected was significantly greater in susceptible interactions. As a consequence, Jordan and DeVay postulated that $OH^{\cdot}$ radical production is favored *in planta* at pH values lower than 5.0. Their electronmicrographs were interpreted as indicating that membrane disruption was closely associated with HR, whereas oxygen radical generation was more a function of pathogenesis. Hence, although reactive oxygen radicals appear to be an attractive basis for HR-related plant cell membrane damage, their precise role remains to be more fully established.

H_2O_2 - the result of the conversion of an elicitor to a second messenger?

In this treatise we attempt to emphasize the earliest of events of HR, specifically those that provide the bases for rapid plant cell membrane leakiness or disorientation. In a series of studies, Low, Apostol, and associates (1986, 1987, 1989a,b) sought to determine the nature of events that occur subsequent to recognition of the elicitor, but prior to the greatly potentiated synthesis of new mRNA molecules reported by Bell and co-workers (1984, 1986). Specifically in question were the events that lead to transduction of a signal molecule to the cytoplasmic side of the membrane, *i.e.*, the conversion of the elicitor molecule to a second messenger. Second messengers effect changes in pH, Ca^{2+} flux, redox reactions, membrane fluidity and, as we shall see, the synthesis of H_2O_2.

Low *et al.* (1986) recognized that signal transduction events in animal cells involve alterations in membrane permeability and intracellular ion concentrations and sought to monitor these in cotton suspension culture cells exposed to a *Verticillium dahliae* elicitor. Using membrane potential-sensitive (oxonol) and ion selective (pyranine) fluorescent dyes they reported changes in membrane permeability within a *few minutes* of elicitor addition. The changes were rapid and steep reductions in fluorescence of either pyranine or oxonol following the addition of elicitor (Fig. 18). The *V. dahliae* elicitor functioned similarly in suspension cultures of tobacco and soybean, suggesting the general applicability of the technique and perhaps reflecting the wide host range of the pathogen.

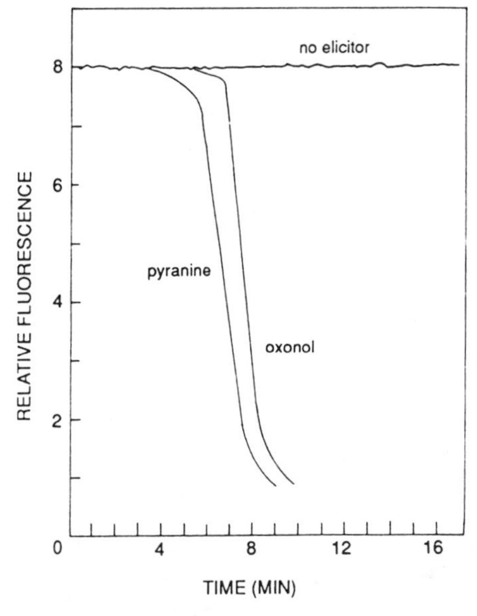

Fig. 18. The effect of an elicitor extract from *Verticillium dahliae* on the fluorescence of cultured cotton cells incubated with either pH-sensitive (pyranine) or membrane-potential-sensitive (oxonol VI) dyes. In the absence of elicitor no change in dye fluorescence was observed. However, upon addition of 10 μl of the elicitor extract, a dramatic decrease in dye fluorescence followed a brief lag period. Two milliliters of a cotton cell suspension were mixed with 2 μl or pyranine or 5 μl of oxonol VI and then 10 μl of either distilled water or a crude elicitor solution of a *V. dahliae* 277 extract was added. Low and Heinstein, 1986.

The Low group (Apostol *et al.*, 1989b) conducted experiments to assess the nature of the metabolic changes occurring in soybean suspension culture cells that accounted for the precipitous drop in fluorescence following their exposure to the *V. dahliae* elicitor. It was clear that the elicitor rapidly stimulated defense activities which were apparent in 0-8 min as a precipitous loss of fluorescence. Membrane associated phenols such as IAA and L-DOPA were as highly sensitive to elicitor-stimulated modification as the dyes were (Fig. 19). Hence, these aromatic compounds also were used as fluorescent probes to monitor the response to the additional of elicitor.

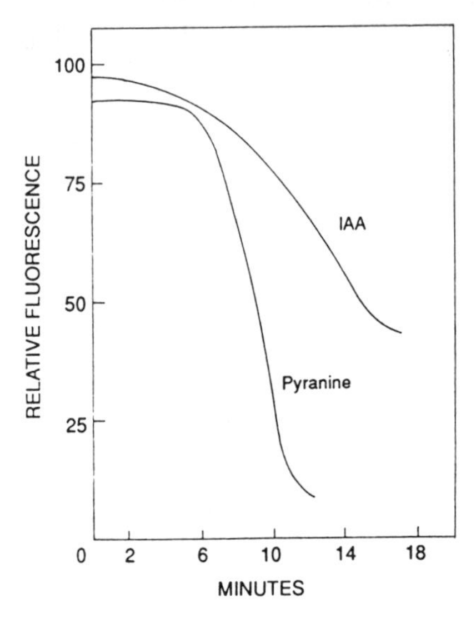

Fig. 19. Effect of addition of elicitor to cultured soybean cells treated with 16 μM IAA. The excitation and emisison wavelengths for monitoring the destruction of IAA were set at 302 and 360 nm, respectively. The fluorescence transition of pyranine in the same cell suspension is provided for comparison. Apostol *et al.*, 1989b.

Since these phenols contained oxidizable functional groups, *the possibility of an elicitor-activated oxidative process was examined* (Apostol *et al.*, 1989b). When soybean cells were treated with superoxide dismutase, mannitol or catalase to destroy elicited $O_2^{\cdot-}$, $OH^{\cdot}$, or H_2O_2, respectively, only catalase obliterated the fluorescence quenching. It was apparent therefore that H_2O_2 was responsible for the fluorescence bleaching either as a direct oxidant or as substrate of an endogenous peroxidase. A comparison of elicitor and 100 μM H_2O_2 on the quenching of fluorescence found both to be nearly instantaneously effective. The Low group interpreted these results to mean that oxidative enzymes were constitutively present and at levels sufficient to synthesize enough H_2O_2 to bleach susceptible compounds. It was shown that known inhibitors of plant peroxidases and electron transport could block both IAA and pyranine elicitor-stimulated oxidation by 55 to 100%.

Of additional interest was the observation (Low and Heinstein, 1986) that *newly regenerated* cell cultures were *virtually unresponsive* to the elicitor with respect to fluorescence quenching. However, *aged cells*, 36-48 h in culture, *responded intensely* and dependably. This observation is particularly reminiscent of the aging required in potato cells and tissue for maximum HR response following inoculation with *P. infestans* (Tomiyama, 1960; Bostock *et al.*, 1983), however, the relationship is not completely clear. Soybean suspension culture cells transferred frequently quenched fluorescent probes at a high level without the addition of elicitors (autoelicited) in the first few day following transfer to fresh media (Apostol *et al.*, 1989a). Those transferred less frequently autoelicited at a lower level and it was shown that these were the cells that dependably responded to elicitor. Sixty to 100 h after transfer soybean suspension cells neither autoelicited, nor responded to elicitor. The level of autoelicitation depended upon plant species as well as transfer schedule. Thus the ability to respond to stress oxidatively depended both upon age and the previous history of environmental stress, a phenomenon previously noted by several investigators studying HR development.

The complete suppression of the initial events of elicitation by catalase suggests that the early phases of phytoalexin synthesis (Apostol *et al.* 1989a) and perhaps HR as well are inhibited by this enzyme. Apostol and coworkers (1989a) also suggested that catalase may act as a inhibitor of signal transduction and that H_2O_2 may perform as a second messenger in phytoalexin development (and perhaps HR as well).

The authors wish to emphasize the fact that the experiments by the Low group detailed here are performed in vitro with suspension culture cells.

The use of the generated H_2O_2 by extracellular peroxidases within minutes of addition of elicitor, hours before phytoalexin detection, suggest that H_2O_2 could be an early defense product. It is known that membrane lipid peroxidation is a consequence of peroxidase activation and thus H_2O_2 may constitute an early feature of HR development (Keppler and Novacky, 1989b). A hypothetical model, proposed by Apostol *et al.* (1989), of the structural relationships of the redox chain that generates H_2O_2 from a cytosolic source of NADH in the plasmalemma and cell wall appears in Fig. 20.

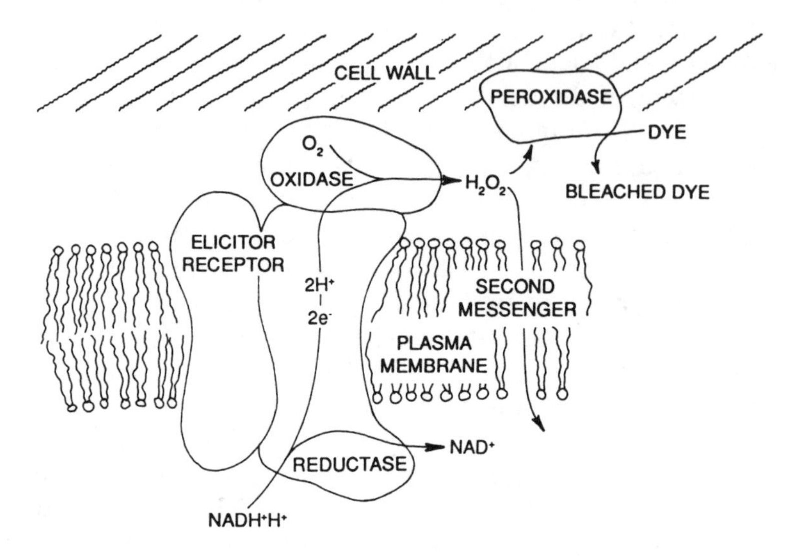

Fig. 20. A hypothetical model of the arrangement of redox components in the plasma membrane/cell wall which may be involved in the oxidative burst. An elicitor receptor is hypothesized to be linked in some manner to a redox chain which consumes NADH/NADPH to generate H_2O_2. The reductase is somewhat arbitrarily oriented across the membrane, since it likely employs cytosolic NADH/NADPH to reduce extracellular O_2 to H_2O_2. The peroxidase may be cell wall associated as suggested by many others. The target of the H_2O_2 in transmembrane signaling could be virtually any component but since mRNA synthesis must occur in the nucleus, the signaling arrow has been sketched pointing toward the cell interior. Based on this model any component which interferes with receptor-reductase coupling or which consumes electrons (*i.e.* an oxidant such as $Fe(CN)_6^{-3}$) or reduces H_2O_2 (e.g. $Fe(CN)_6^{-4}$ or which blocks the oxidase (e.g. KCN) would be expected to inhibit both the bleaching and transmembrane signaling functions of the oxidative burst. Apostol *et al.*, 1989.

Schwacke and Hager (1992), have also presented evidence that cell wall components of ectomycorrhizal fungi *Aminita muscaria* and *Hebeloma crustilinforme* and the spruce pathogen *Heterobasidium annosum* at comparatively low concentrations elicited a release of active oxygen species from cultured spruce cells, *Picea abies*. Using the chemiluminescent luminol, the oxygen burst was detected in 6 min and it peaked after 24 min in the reaction mixture of spruce cells and elicitor. The levels of active oxygen detected were elicitor dose dependent. The nature of the oxygen species being evolved in this burst was presumed to be H_2O_2 as luminescence could be nearly totally suppressed by catalase. However, in the presence of added SOD the levels of detectable H_2O_2 were increased, suggesting that some superoxide was also being formed. Since both oxygen species may have been present it is possible that a third form, the hydroxyl radical, may have been as well.

Of additional significance was the finding that Ca^{2+} is integral to the development of the oxidative burst, as depletion of this cation from the medium diminished the production of H_2O_2. Similarly, the protein-kinase inhibitor staurosporine inhibited the oxidative burst, suggesting the probable participation of a phosphorylative event in the process. Both Ca^{2+} and protein kinases have been shown to be integral to respiratory bursts of mammalian neutrophils. The early studies of Tomiyama and Stahmann (1964) alluded to the participation of protein kinase as a precondition in the formation of active oxygen.

A sequence of events in the elicitation of HR may be presumed from the forgoing data. Upon elicitation, Ca^{2+} acts as a transducing element and in concert with protein kinase phosphorylase activates a cytosolic protein. This protein may bind to membrane bound NADPH oxidase which in turn generates superoxide by reducing molecular oxygen.

Site of action and fate of the elicitor *in planta*

The Low group, Horn *et al.*, 1989, pondered the fate of the fungal elicitors of defense against pathogenic attack. Plant hormones are comparatively small compounds that readily enter the cell and reach their target sites. Further, they are generally controlled by specific enzymes or complexing molecules that limit their activity *in situ.* Whereas receptor-mediated endocytosis of animal cells offers a process of clearing extracellular ligands (elicitors), no such mechanism has been proferred for plant cells. Specifically, how are receptor sites for *V. dahliae* elicitor on soybean cells "neutralized" following elicitation? The question actually posed is "what happens to polysaccharide and glycoprotein elicitors that bind to plant cell surfaces" and are seemingly too large to penetrate the cell and what turns them off following elicitation?

Using fluorescein- and ^{125}I-labelled elicitors, it was shown that the fluorescein isothiocyanate labelled *V. dahliae* elicitor and similarly a fluorescein thiosemicarbazide-labelled polysaccharide from pectin were rapidly, within 1-4 and 2 h respectively, internalized by soybean cells. The receptor-mediated uptake of elicitor was found to be ligand-specific, and internalization of the elicitors was an energy-dependent, temperature-sensitive, receptor-mediated process. It was found that random uptake of extracellular non-ligand molecules proceeds at a much slower rate than elicitor endocytosis. It was also shown that the endocytosed elicitors were transported into the cell *in toto*, could be extracted, and were subsequently identified as intact.

With ^{125}I-labelled polygalacturonic acid, elicitor uptake proceeded at 10^6 molecules/cell/min. The fluorescent probe elicitors initially associated with the cell surface and then accumulated *in the vacuole.* The internal sequestration took place well after the release of H_2O_2, which occured within 5 min of cell exposure to the elicitor. *Hence, receptor-mediated endocytosis, now shown clearly to occur in plant cells may well be the process by which elicitor provoked stimuli are "switched" off once the signal has been transduced.* Thus, once the plant cell has detected the presence of the pathogen and it has assembled its defenses, the ligand is cleared from the cell surface and metabolism is returned to a ground state to continue normal anabolic processes.

It would also appear that certain large *ligands* (*e.g.* large polymer molecules) may have specific channels through which they penetrate plant cell walls and plasmamembranes. *This finding appears to broaden the scope of large organic molecules that may penetrate the plant cell wall and enter the cell per se.*

The ontogeny of H_2O_2

In following the ontogeny of the burst of H_2O_2 detected by the Low group (1989b) and recognizing the two distinct periods of $O_2^{\cdot-}$ generation noted earlier by Chai and Doke (1987), we were attracted to the report by Elstner and Heupel (1976). Their study revealed that isolated cell walls of horseradish (*Amoracia lapathifolia*) in the presence of $MnCl_2$ catalyze the production of H_2O_2 at the expense of either NADPH or NADH. Further, the reaction is stimulated by catalytic amounts (3×10^{-5}M) of *p*-coumaric acid. Related aromatic compounds with 4-hydroxyl and 3-methoxyl substituents, such as ferulic acid and coniferyl alcohol were particularly stimulatory. Hence, these studies provided evidence that a plant-cell-wall bound peroxidase produced both H_2O_2 and a superoxide free radical ion as intermediates of a complex chain reaction. The ubiquity of the enzymes and the above aromatic substrates in plant cell walls in general and their utility in lignification processes suggests that they may be precursors for the generation of H_2O_2 detected by the Low and Doke groups and are probably readily available in the cell walls of most plant systems. For example, Ebel *et al.* (1984), ascribed phytoalexin synthesis in soybean cells exposed to a glucan elicitor of *P. megasperma* f. sp. *glycinea* to increases in mRNA that correlated with PAL and CHS. However, they noted that even unelicited cells contained significant levels of both enzymes.

It is of interest to consider the location of the peroxidase enzymes that control the elaboration of H_2O_2. According to Goff (1975) they are not only in the cell walls *per se*, but are also in the plasmalemma, Golgi vesicles, endoplasmic reticulum and on membrane bound ribosomes. Hence peroxidase is a secretory protein; and according to Fry (1986) possibly a regulatory factor controlling plant cell wall rigidity. Ostensibly this enzyme mediates the lignification process, as well as associated reactive oxygen molecules (Goodman, Király, and Wood, 1986) and the structural development of the plant cell wall.

Elstner and Heupel (1976) have reported that the secretion of peroxidase requires the presence of Ca^{2+} and secretion of the enzyme can be detected in 30s after the addition of the ion (Sticher *et al.*, 1981). In fact release of the enzyme can be observed following increases in Ca^{2+} as low as 10^{-6} M. Calcium is believed to contribute to the structural stability of peroxidase as well as its enzymatic activity, by binding the enzyme to plant membranes or wall components (Sticher *et al.*, 1981). The Ca^{2+} enters the cytosol, causing peroxidase to be adsorbed to a vesicular membrane and in this form is carried to the plasmalemma where it is exocytosed. The precise mechanism of secretion is still unknown, however, proposed interactions are with auxin, cyclic AMP, Ca^{2+}/calmodulin and Ca^{2+}-ATPase. Inhibition of peroxidase release by NaN, in the presence of Ca^{2+} indicates that release is metabolically controlled, perhaps by calmodulin. The retention of peroxidase in frozen cell wall, though exposed to Ca^{2+}, suggests that Ca^{2+} releases the enzyme by means of a specific process. The rate at which peroxidase is released (Sticher *et al.,* 1981) suggests that it is a preexisting molecule that accumulates in an inactive form and then may be released in response to Ca^{2+}.

According to Castillo (1986), the stimulation of peroxidase activity is a stress related phenomenon and is indicative of a controlled general increase in oxidative processes. Castillo recognized both *basic* and *acidic* peroxidases, see Fig. 21; the former are released from the cytosol into the cell wall or free space as a consequence of stress and ionic flux and appear rapidly. The acidic form of the enzyme is activated later and is probably involved in lignification and perhaps phytoalexin synthesis. With respect to HR, it would seem that the basic peroxidases may be the effectors of H_2O_2 and $O_2^{\cdot -}$ accumulation.

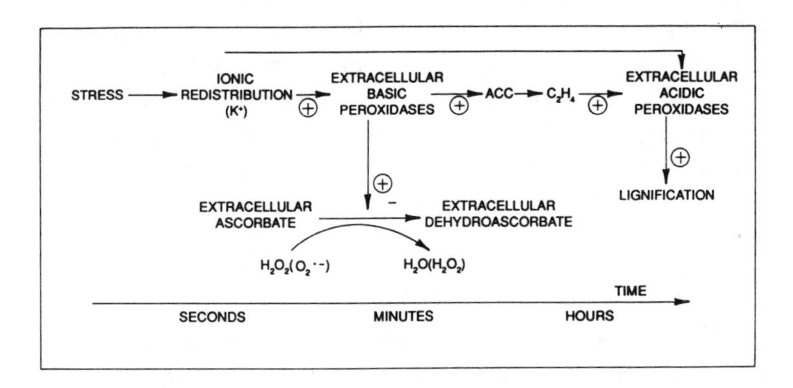

Fig. 21. Possible pathway of peroxidase reactions in response to oxidant stress. Castillo, 1986.

The report by Castillo suggests that an increase in cytoplasmic Ca^{2+} is required for peroxidase secretion. The newly secreted peroxidase is activated by the comparatively high concentration of Ca^{2+} in intercellular space and in so called "free space" (plant cell wall). Apparently stress causes Ca^{2+} to increase in the cytosol which causes the peroxidase to diffuse out into the cell wall where it is activated by wall Ca^{2+}. The influx of Ca^{2+} into the cytosol may be in response to an efflux of K^+ that has been recorded by Atkinson (1989, 1990) in bacterially induced HR.

Returning to the "burst" of H_2O_2 observed by the Low group, that is generated in the presence of NADH, (Gross *et al.*, 1977), and is stimulated by monophenols like p-coumaric acid, ferulic acid and coniferyl alcohol (Elstner, 1982): the NADH was shown to be provided by a bound malate dehydrogenase in horseradish cell wall. *In this system the enzyme could polymerize monophenols, yielding an insoluble dehydrogenated polymer i.e., condensed tannins resulting in brown pigment development* (Srivastava and Van Huystee, 1977). The nucleotide NADH may in fact be oxidized by two different mechanisms, the first involving a monophenol, Mn^{2+} and a superoxide radical in a reaction that is insensitive to SOD. A second mechanism appears to be dependent upon free $O_2^{\cdot-}$ that is generated along with H_2O_2, and a malate dehydrogenase -NADH complex.

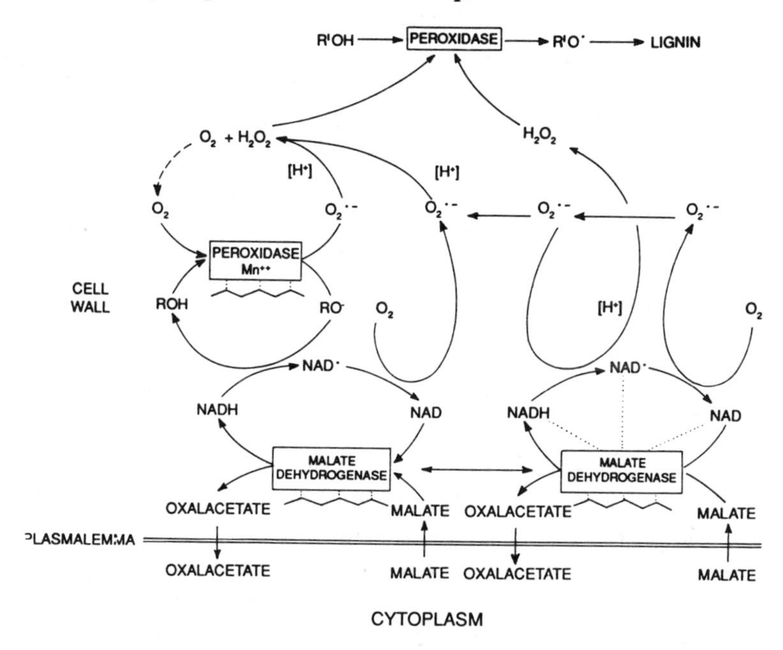

Fig. 22. Proposed scheme of reactions generating hydrogen peroxide in higher plant cell walls. Sum of reactions: 2 malate +2 O_2 →2 oxalacelate +2 H_2O_2. Gross *et al.*, 1977.

It would appear that 4-hydroxylated cinnamyl alcohols or their acids, which are plentiful in plant cell walls, could readily serve as H^+ donors when H_2O_2 and $O_2^{\cdot-}$ are formed and NADH is, conceivably, the immediate electron donor. Although monophenols are competent electron donors, there is some evidence that malic acid is the principal electron donor. A hypothetical scheme for reactions generating H_2O_2 and $O_2^{\cdot-}$ is shown in Fig. 22 (Gross *et al.*, 1977).

From the foregoing it would seem that there is reasonable evidence suggesting that reactive oxygen radicals play a role in HR development. Both $O_2^{\cdot-}$ and H_2O_2 generation may be involved and some factors initiating and potentiating their activity such as Ca^{2+} and peroxidase seem likely in this regard. A hypothetical scenario for the *process* of the induction and development of HR will be detailed in the concluding "Summary" section.

Expression of HR and pathogenesis genes

It is apparent that HR is an ubiquitous and persistent phenomenon in the development of resistance in plants to some fungal, bacterial and viral pathogens. In most instances where HR has been studied it is clear that HR proscribes a gene-for-gene relationship. From a biochemical point of view, the resistance gene or genes involved code for a receptor molecule that effects HR upon interaction with an avirulence gene of the pathogen which in turn codes for an elicitor or initiator of the HR resistance phenomenon.

Genetic manipulation of the fungal genome

The possibility of manipulating the genomes of bacteria and viruses has made it possible to expose specific genes and in some instances their products. This has been more difficult to accomplish with the fungi. The obligate fungi have been particularly recalcitrant, specifically because of the contraints imposed by their cultural conditions. Although DNA transformations have been accomplished with some biotrophic and necrotrophic fungi, *i.e. Cladosporium fulvum, Rhynchosporium secalis* and *Magnaporthe grisea* (de Wit, 1992), lack of really efficient transformation systems has seriously delayed progress. As de Wit has indicated, "eventually one has to prove the role of a putative resistance or avirulence gene by transferring it back into the host or pathogen of interest".

The most promising genetic studies concerning the fungi with respect to genetic manipulation have been with the *C.*

fulvum-tomato model developed in the laboratory of Pierre de Wit. This system is comparatively simple as it uses intercellular fluids (de Wit and Spikman, 1982) of the host to study fungal proteins that are constitutively produced or specifically induced during colonization of the plant. The proteins influence basic compatibility (susceptibility) or elicitor activity inducing HR (de Wit and Oliver, 1989). One of the elicitors, the putative product of avirulence gene *avr9*, interacts specifically with resistance gene *Cf9*, which has been purified and sequenced by Scholtens-Toma and de Wit, 1988). The affect of the elicitor was not detected in compatible interactions (ostensibly because there were no receptors) which lacked a functional *Cf9* gene, indicating the absence of the putative necrogenic *avr9* gene product. It was subsequently shown that races of the fungus that are virulent on tomato genotypes carrying the *Cf9* resistance gene have no DNA homologous to the *avr9* gene, *the first fungal avirulence gene that fits the gene-for-gene concept* (Van den Ackerveken *et al.*, 1991 and Van Kan *et al.* 1991).

A subsequent publication by Van den Ackerveken *et al.* (1993), reported that whereas *C. fulvum* produces no *avr9* elicitor *in vitro,* high levels of *avr9* gene expression is accomplished under the control of a heterologous *gpd* promoter of *Aspergillus nidulans*. Transformants of *C. fulvum* harboring the *gpd* promoter-*avr9* fusion produced and excreted a set of *avr9* race specific elicitors *in vitro.* Culture filtrates therefrom induce HR in tomato plants carrying the *Cf9* resistance gene.

Purification of the culture filtrates revealed a series of peptides not detected earlier by Scholtens-Toma and de Wit, (1988), because of differences in the extraction and purification procedures. However, the most recent study by Van den Ackerveken *et al.* (1993), revealed that plant factors mediate the processing of a 34 amino acid (aa) peptide into the 28 aa form reported earlier by Scholtens-Toma and de Wit.

In brief, these latest data propose the following model for the processing of *avr9*. The precursor protein encoded by the *avr9* gene is 62aa's in length (Van Kan, 1991), whereas the elicitor peptide isolated from *C. fulvum*-infected tomato has the C-terminal 28 aa's of the precursor protein detected by Schlotens-Toma and de Wit. Clevage of the 63aa is accomplished by extra cellular proteases that yield 32 and 34aa species. These are further altered by plant factors that produce the "mature" 28aa race-specific HR elicitor peptide. With ^{125}I-labeled AVR9 peptides as ligands, receptor-binding studies are in progress. These suggest that a "hypothetical receptor on the resistant host plant might be the primary product of the complimentary resistance gene *Cf9*" which recognizes the AVR9 peptide elicitor. Hence, we now have been made aware of a 28aa peptide, coded for by the *avr9* gene of *C. fulvum,* that induces HR in tomato that carries the *Cf9* resistance gene. Additionally, the peptide sythesized by the fungus is further processed by plant factors that modify and stabilize this HR elicitor.

The situation regarding the genetic control of the fatty acid elecitors AA and EPA is at this writing without resolution. It has been suggested that AA may be released more rapidly in the case of the resistant than the susceptible reaction (Bostock, personal communication). In this instance, the resistant host reaches the HR gene activation threshold level more quickly and thus precludes the expression of those genes that generate pathogenesis.

In support of this contention, Ricker and Bostock (1992) have demonstrated, with radiolabeled *P. infestans* sporangia used as a foliar inoculum, that AA and or its metabolites are released three hours earlier in the resistant reaction (nine *vs.* twelve h). The metabolites of AA diffuse within the developing lesions and are able to act as signal molecules to the surrounding host cells. In fact, radioactivity (ostensibly AA or its metabolites) was distributed in cells several layers distant from the site of inoculation and from fungal infection structures. Of additional importance was that this triggering of HR was caused by the sporangia *per se,* and did not require extensive pathogen colonization. It was also apparent that the signal molecules have at least limited systemicity within the inoculation sites during the early phases of the infection process (perhaps the first nine hours after inoculation of potato foliar tissue with *P. infestans* sporangia).

THE VIRUS-INDUCED HYPERSENSITIVE REACTION

Introduction

The development of necrotic leaf lesions caused by viruses in a large number of hosts, following natural or artificial inoculation has long been the subject of intense study (Holmes, 1929; Weintraub and co-workers, 1958, 1979). It became apparent that the local lesion response to virus infection reflects the plant genotype, *e.g.* the tobacco N or N' gene, which may be modified by temperature, light, inoculum dosage, and inoculation efficiency. On close inspection of local lesion formation it was also clear that the phenomenon bore important similarities to what Stakman (1915) had called the hypersensitive response (HR).

The local lesion response is in fact the development of foliar necrosis and the confinement of the virus in or very near the necrotic zone. K.R. Wood (in Goodman *et al.*, 1986) has indicated that hypersensitivity, as it applies to virus-infected tissue, implies *a priori* that the infected tissue becomes necrotic. However, there are reports in the literature that necrosis is not always accompanied by virus localization or that virus localization may occur without necrosis. In fact if HR development proceeds in total darkness, necrosis can be obviated (Peever & Higgins, 1989). Additionally, it has been frequently demonstrated that virus infection may become systemic following the development of necrosis. From the foregoing, there can be little doubt that the relationship of necrosis to either resistance or virus localization is in need of further clarification. More specifically, it would seem that the relationship of HR to resistance, as exemplified by necrosis and subsequent virus localization, requires additional intensive study and that HR must be more precisely proscribed.

What is virus-induced HR?

Some insight into this question may be obtained from a more precise characterization of the necrosis that reflects HR. Specifically: is virus replication implicit; how many host cells are affected; what is the time span between inoculation and host cell collapse? Finally, are necrosis and localization of the virus requisite features of HR or subsequent events?

If one defines HR as rapid host cell death, with the eventual necrotization of the dead or dying cell, then it would seem the rate at which cells die as a consequence of exposure to virus is crucial. For example, if relatively few host cells are inoculated and die before extensive virus replication and transcellular migration occurs, both necrosis and localization may develop rapidly. However, if cell death is slow, permitting both replication and systemic movement, it is apparent that although necrosis develops, it may be at a rate too slow to prevent local lesion size increase and systemic spread of the virus. Analogously, the observation that virus localization occurs in the absence of necrosis suggests that other factors may be responsible for localization, for example, callose deposition.

On the other hand, TMV-induced lesions in the N' gene-containing White Burley tobacco continue to grow, and the size of lesion that ultimately develops appears correlated with the thermal stability of the coat protein of the virus strain (Fraser, 1983). This implies that necrosis reflects an interaction between host cell and virus coat protein.

That virus-induced HR is cytologically disruptive has been recorded ultrastructurally for example, in *N. glutinosa* inoculated with TMV by Weintraub and Ragetli (1964b). It is also apparent that HR is a cytotoxic response that develops quickly in host cells sensitive to avirulent gene products of viral plant pathogens. Details of these and other observations will be discussed in subsequent sections of this monograph. However, it is clear that analogous host cell multi-organellar membrane destruction caused by some viral pathogens parallels HR induced by plant pathogenic fungi and bacteria.

As we mentioned at the outset, where our definition of HR was presented (see page 7), the plant is able to mount a series of diverse and complex resistance reactions. We recognize that as a consequence of virus infection the deposition of polymers such as callose, glycoproteins and lignin may occur which contribute to or are singular detectable features of resistance. The synthesis of specific antiviral factors and localizing envelopments may also contribute to the suppression of the virus pathogen. However, in this monograph we are concerned only with the rapid development of host cell death and necrosis which is known conventionaly as HR or local lesion development.

Infection process of TMV in tobacco resulting in HR (local lesion)

Van Loon (1987) has written that, since virus cannot be isolated from tissue more than a few cell layers from the edge of a mature lesion, the lesion is an accurate measure of virus spread. It is also an indication of the intensity and rapidity of host cell collapse. The implications of local lesion necrosis to HR and resistance are immense. However, Dunigan, Golemboski and Zaitlin (1987) have stated that the hypersensitive response (HR) (local lesion development), probably the single most studied phenomenon in plant virology is, nonetheless, still not completely understood in many respects.

It would appear that a parallel exists in the infection processes exhibited by race 0 of *P. infestans* in potato cvs. Irish Cobbler and Rishiri and the U_1 race of TMV in Samsun and Samsun NN tobacco (Tomiyama, 1960). There is good evidence that in the incompatible combination *P. infestans* initiates the HR process once it has penetrated the host cell wall. Germination and the process of wall penetration are completed at the same rate by incompatible and compatible races of the fungus. Analogously, the inoculation of Samsun NN or Samsun with the HR-inducing TMV strain U_1 causes virus replication to proceed at the same rate in both cultivars for the first 30 h following inoculation (Konate and Fritig, 1984). It is only after that time that virus replication in the HR^+ cv. Samsun NN is significantly reduced. It is not known whether virus replication *per se* or just virus RNA penetration of a Samsun NN host cell is sufficient to initiate the sequence of events that culminate in discreet local lesion development and the reduction in virus replication. According to Fritig *et al.* (1987) HR does not preexist in the host cell, but is induced by infection and the N gene product and the avirulence gene product of U_1 react for 30 - 36 h in order for HR necrosis and host cell death to occur.

In a review of the early moments of virus initiated plant infection Siegal (1966) concluded that "a single particle is sufficient to convert an infectible site to an infective center". The rapid development of a small local lesion also implies that some additional host cells have been infected and that there was migration of virus from the initial infectible site to a number of adjacent cells.

The rate of initial TMV inoculum replication in epidermal cells of tobacco and subsequent migration to new infection sites has been determined. In 1963 Fry and Matthews performed a number of classic experiments indicating that newly made virus from the initial inoculum appears in about 7 h. They also reported that migration of virus particles to mesophyll cells required an additional 4 h. (See Fig. 23 and 24)

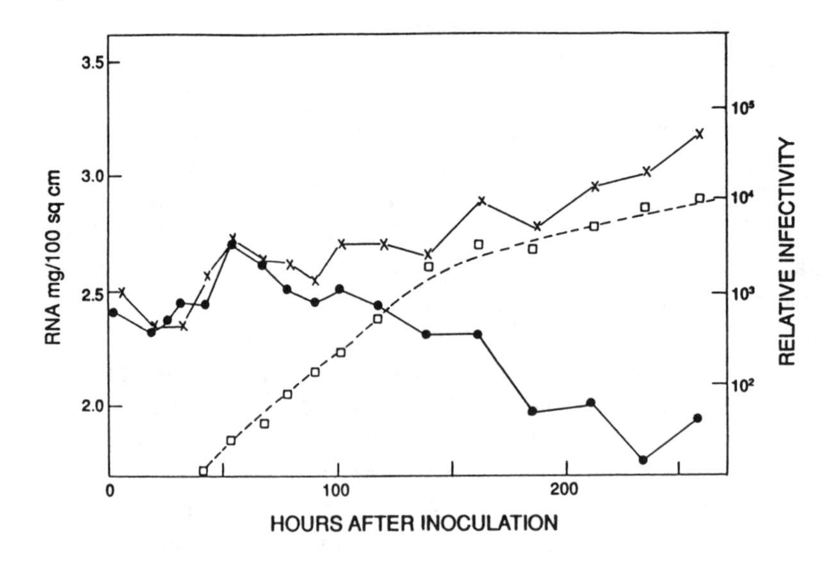

Fig. 23 Total RNA content of TMV-infected (X) and uninfected (•) half-leaf samples following inoculation. Relative infectivity (□) of extracts of infected half-leaves. Fry and Matthews, 1963.

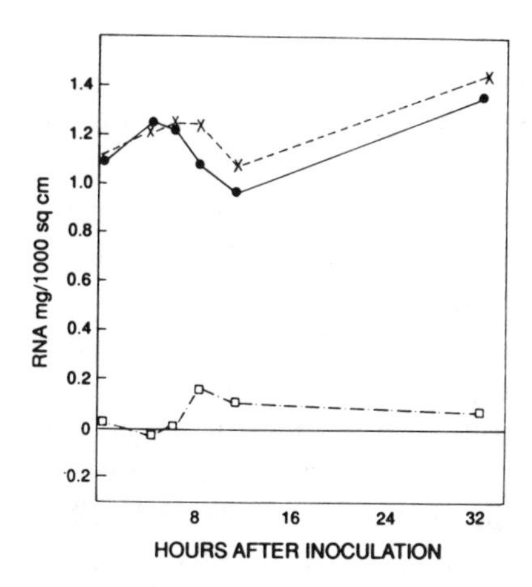

Fig. 24. Total RNA content of TMV-infected (X) and uninfected (O) epidermis following inoculation. □ = Difference, infected minus uninfected. Fry and Matthews, 1963.

It is still not known whether the single local lesion reflects the progeny from a single infectible cell as considered by Siegal (1966). In a study by Konate and Fritig (1984), the exact locations of a large number of microinoculation sites, 0.02 - 0.05 mm^2, were marked and virus titers were determined over a time course. The procedure engendered the development of a single local lesion with 90% efficiency when a saturation inoculum dose of 500 μg/ml was used. This technique permitted the discrimination of the original inoculum/lesion from the progeny of the inoculum by using radiolabelled uridine (Tab. 2). Hence, the replication and cell to cell spread was readily calculable in more than 3,000 individual microinoculation sites.

Tab. 2. Use of the radiochemical method to distinguish between the original inoculum and the progeny developing therefrom. Konate and Fritig, 1984.

	Inoculation site no.						
	1	2	3	4	5	6	7
Virus amount (ng/site)	0.20	0.20	0.36	0.40	0.80	1.20	2.60
Radioactivity (cpm/site)	4	3	7	5	605	790	2010

The microinoculation procedure was performed on attached *N. tabacum* cv. Samsun leaves with a purified suspension of 500 μg/ml of TMV U_1. One day later the zones surrounding the inoculation points were each infiltrated with 20 μCi of [^{3}H]-uridine. After a further 8 h of incubation 5-mm disks centered on the inoculation sites were punched out and assayed for virus titers and radioactivity in the virus.

It is apparent, with respect to the elicitation of HR, that the question whether translocation and replication are requirements for induction remains to be answered. Specifically, is cell death the reflection of the influence of TMV particle(s) in the penetrated cell; or must a number of cells be penetrated to manifest the small lesion; or do uninfected cells surrounding the penetrated cell influence and/or cause the death of the penetrated cell? Since lesions visible to the naked eye do develop, it would appear that some translocation from initially infected cells does occur. Furthermore, since additional cells are penetrated from the initially low number that are infected, virus replication is also implicit in local lesion-HR cell death. The hypothesis that replication is implicit in HR induction, could be tested by inoculations with a defective RNA (Zaitlin, personal communication).

Since necrosis does not occur in TMV-infected isolated plant cells it seems valid to concur with Otsuki *et al.* (1972) that the process of necrotization is "based on an interaction of infected cells with cells of the surrounding tissues which may or may not be infected." This contention seems to be indirectly supported by the report of Beachy and Murakishi (1971) that N gene-containing cultivars, NN Samsun and NN Burley, both form necrotic lesions in callus cultures (cell aggregates). However, no evidence has been presented indicating HR necrosis development in *isolated* N gene-containing cells infected with TMV.

In a slightly different scenario reported by De Laat and Van Loon (1983b), it was shown that by alternately shifting temperature of tobacco inoculated with TMV from 20° to 30° C, lesion size in Samsun NN inoculated with TMV WU_1 reflected a balance between virus replication and spread, and the efficiency of N gene resistance (Fig. 25). N gene resistance is influenced dramatically by temperature and additionally by leaf age and light (photosynthesis). It would seem then, that N gene activation or the triggering of HR at 20° C occurs, according to data from De Laat and Van Loon (1983b), around 15 - 18 h after inoculation. Ostensibly, this is the period during which the essential virus replication occurs. Accordingly, the N gene and the TMV genome must interact for an additional 6 to 9 h "for the initiation of a series of events that culminates in a burst of ethylene production and the formation of necrotic local lesions 24 h later". De Laat and Van Loon (1983b) also contend that N gene activity for a minimum of 3 h is necessary for transcription and translation of the respective genomes to take place. In total then ~30-33 h are required for earliest symptoms to develop.

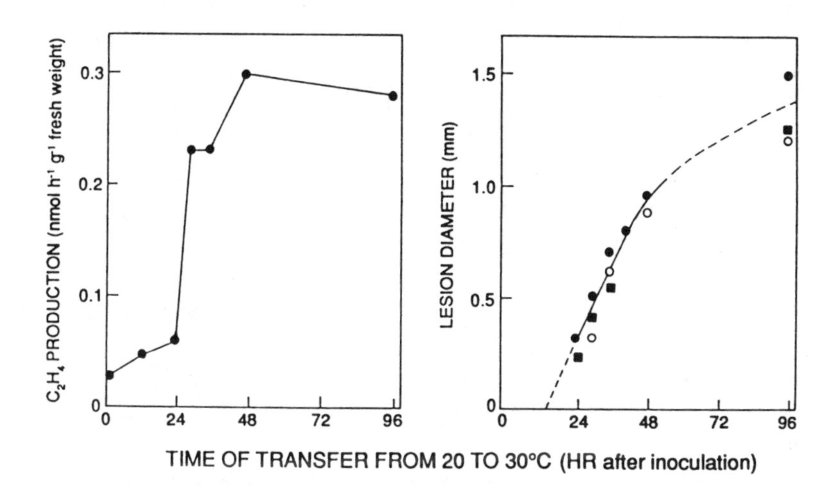

Fig. 25. Effect of a shift in temperature from 20° to 30° C at successive intervals after inoculation of Samsun NN leaves with TMV on (left) ethylene production between 36 and 48 h after inoculation, and (right) lesion size measured four days after inoculation ●, ■, □; data from three different experiments, lesions visible 42 h after inoculation. De Laat and Van Loon, 1983b.

Portals of entry for virus particles or RNA

From the data currently at hand it would appear that either virus replication or contact of virus particles with the host cytoskeleton is necessary for HR development. In reconstructing the events that may occur following inoculation we may presume, at least in tobacco, that microlesions permit entry and migration of the virus particle or viral RNA into the host cytosol. Studies by Nishiguchi *et al.* (1986) indicated that a technique known as electroporation assisted in the delivery of plant viral RNA into tobacco protoplasts. The procedure utilizes an electrical impulse to cause reversible breakdown of plasmamembranes. With this technique, a concentration of 5-50 μg/ml of TMV or CMV RNA permitted 90% of the protoplasts in the suspension culture to become infected. In contrast, more conventional procedures with an array of adjuvants infected about 20% of the protoplasts.

The electrical impulse-induced modification of the membrane which produced the micropores was presumed the result of membrane compression by the high field intensity of the impulse. Pore density was estimated to be $10^7/cm^2$. The pore size to accommodate virus passage, based on the dimensions of a TMV particle, 300x18 nm, should be larger than 30 nm. However, particle or RNA molecular size (configuration) are apparently not the only constraints for viral passage through the plasmalemma.

Since protoplasts have long been shown to be infected by virus particles or viral RNA without the potentiation by electroporation, it is possible that there are natural pores in the plasmamembrane of tobacco protoplasts. One may presume that abrasion of tobacco leaf surfaces exposes the plant cell membrane at pore sites thus facilitating entry. Perhaps in nature the pore frequency is not as great as that induced by the electrical impulse used in effecting electroporation. Of course there are other scenarios that may explain transmembrane passage, for example, pore widening due to stimulation of membrane fluidity by the virus particle *per se*.

A further comment on the size of microlesions that permit virus particle entry: they must ultimately seal in order that cells may live and support replication. In this regard studies concerned with membrane potential measurements provide some idea of the size of lesion that rapidly seals after perforation. (A. Novacky, personal communication). In studies of E_m lesions 0.5 μm or larger are commonly made and seal rapidly. It is likely that many microlesions that are made by carborundum exceed the limit of wound healing; hence, even if virus entry is accomplished, replication and subsequent migration do not follow.

Role of temperature on replication and necrosis

The direct control of TMV replication and the expression of the N gene in necrotic-responding varieties of tobacco was critically evaluated by Otsuki, Shimomura and Takebe (1972). In experiments with "necrotic" cultivars Samsun NN and Xanthi nc and "systemic" cultivars Samsun and Xanthi it was demonstrated that the TMV OM strain replicated at 28° C at the same rate in necrotic and systemic cultivars and lesions were detected in neither cultivar type. At 22° C the rate of replication in both types was slower but almost identical for the initial 24 h when pinpoint lesions became visible and multiplication decreased in Samsun NN. Although replication was reduced in Samsun NN (see Fig. 26) lesions continued to enlarge and virus titer increased during the 96 h incubation of the inoculated leaf disks. It appeared that lesion formation *per se* restricts virus replication.

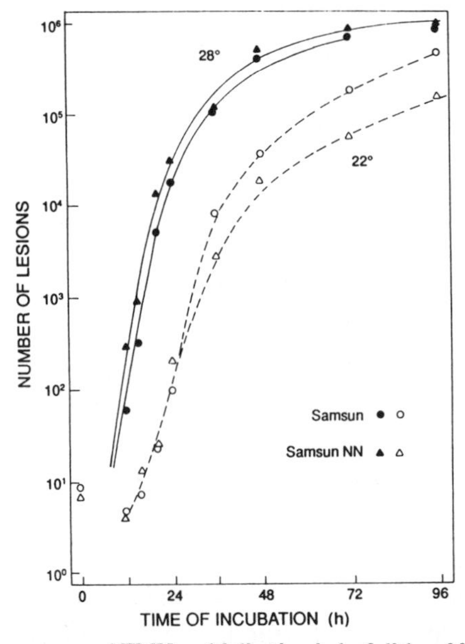

Fig. 26. Time course of TMV multiplication in leaf disks of *Nicotiana tabacum* cultivars Samsun and Samsun NN. Otsuki *et al.*, 1972.

This study was continued using protoplasts of the aforementioned cultivars. At both 22° C and 28° C replication was nearly identical in the two cultivar types (see Fig. 27) although there was a slight lag in replication at the lower temperature. Since the cultivars carrying the N gene did not express necrosis at 28° C in leaf disks, nor in protoplasts at either 28 or 22° C it was concluded that the N gene has no influence on the inherent capacity of individual tobacco cells to synthesize TMV.

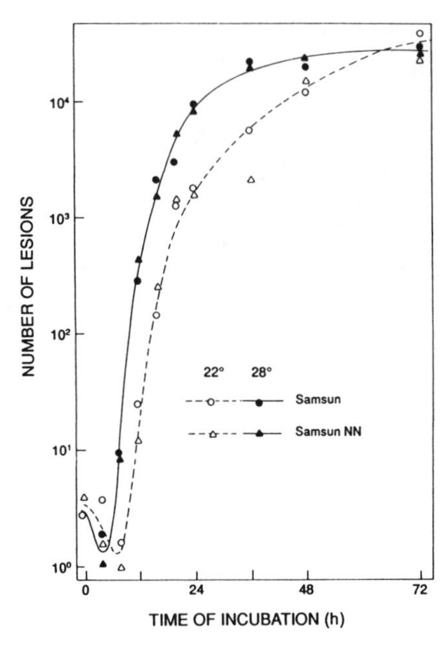

Fig. 27. Growth curve of TMV in protoplasts of *Nicotiana tabacum* cultivars. Samsun and Samsun NN. Otsuki *et al.*, 1972.

Spread of the virus at 22° C in either Samsun NN or Samsun took place during the early phase of infection with analogous speed. Hence, the N gene did not appear to directly influence the cell-to-cell movement of the virus, an observation that has subsequently been validated by Padgett and Beachy (1993), (see page 115).

However, movement and replication were both slowed as soon as necrotic lesions formed. These data may be interpreted as providing evidence that the inhibition of virus replication and spread are separable phenomena from those that elicit cell collapse and necrosis. It also seems apparent that neither replication nor virus spread are restricted by those early events (prior to 24 h after infection) that evoke host cell collapse and necrosis.

Membrane damage and electrolyte leakage

A careful look at cell collapse may assist in understanding what is actually happening. The sequence of near simultaneous development of intense electrolyte leakage and symptoms prompted Ohashi and Shimomura (1976) to conclude that they are directly and temporally related. There is clearly a loss of cell turgor and an efflux of electrolyte, primarily K^+ ions. This is a characteristic of HR-associated membrane perturbation whether caused by incompatible bacterium, fungus or virus. It reflects membrane damage and the possible target, specifically as we shall see, a lipid component of cell organellar membranes.

Weststeijn (1978) presented evidence that TMV-infected Xanthi nc tobacco (N gene-containing), exhibited electrolyte leakage some 7 h prior to the appearance of initial symptoms and 1 h before cell collapse could be detected. It was apparent that continued electrolyte leakage is a precursor to necrogenesis. The reports by Weststeijn (1978) and later by Pennazio and Sapetti (1982) suggest membrane damage precedes by several hours the first visual indications of foliar damage.

Weststeijn (1978) critically studied HR-induced membrane damage and consequent electrolyte leakage from TMV-inoculated *N. tabacum* cv. Xanthi nc. With a heavy inoculum (2-4 mg/ml), which caused eventual inoculated leaf-half collapse, initial symptoms of tiny depressions at leaf undersides appeared in 28 to 34 h near margins. Electrolyte leakage was measured at hourly intervals starting 16 h after inoculation. Although increased leakage was apparent from 18 h after inoculation, it became statistically significant after 21 h. Symptoms developed at 28 h after inoculation at the time when electrolyte leakage began to intensify. Four hours later, leakage intensified further suggesting total loss of osmotic control by cell plasma membranes. These data, shown in Fig. 28, suggest that membrane damage occurs gradually from 18 to 28 h after inoculation. Inoculations of the systemic host Samsun revealed no increases in electrolyte leakage following inoculation.

Exploiting the fact that TMV-induced HR-necrosis of Xanthi nc is a temperature-dependent phenomenon, Weststeijn conducted a classic experiment using a comparatively low inoculum level (0.1 μg/ml) to produce just a few lesions per leaf. Inoculated plants were kept at 22° C until tiny lesions formed. The plants were then moved to 32° C for 3-4 days where local lesion development was precluded but systemic spread continued unimpeded. It is probable that during this period cell-to-cell spread was intense and large numbers of cells adjacent to the

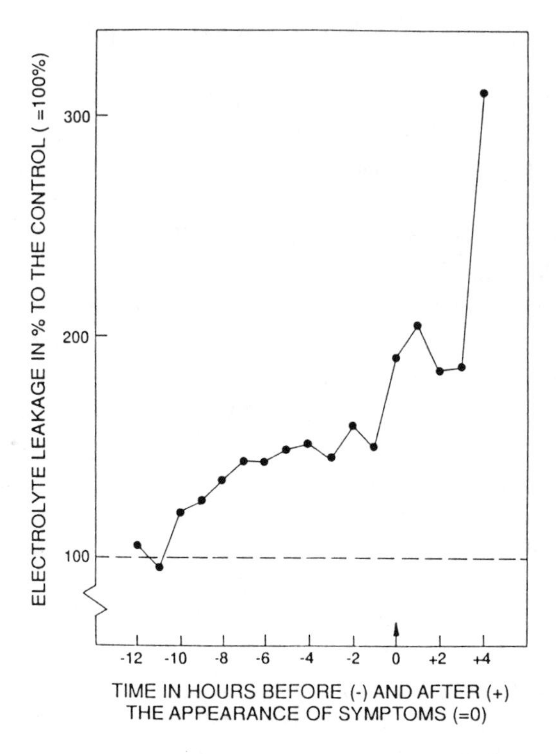

Fig. 28. Electrolyte leakage from TMV-infected *Xanthi* nc leaf disks as a percentage of electrolyte leakage from control leaf disks. Values are means of 3 to 7 experiments. Differences between infected and control are statistically significant from 7 h before symptoms appear onwards. Weststeijn, 1978.

initially infected cells also became infected. The plants were then again transferred back to 22° C and the electrolyte leakage was monitored at hourly intervals from 2 hours after the transfer. Watersoaking around initial primary lesions became apparent 7.5 - 8 h after transfer and electrolyte leakage was detected 1-2 h prior to this development. Weststeijn (1978) suggested that the "first onset of increased membrane permeability may occur earlier than 2 hr before cell collapse". A similar conclusion was reached by Pennazio and Sapetti (1982). *Their data revealed that damage to the membrane is slowly cumulative over a 10 h period, and implied a continuing remedial or wound repair phenomenon that is finally overcome, resulting in irreparable damage.*

The rapid virus replication that one assumes occurred when inoculated plants were transferred to and maintained at 32° C and the subsequent intense electrolyte leakage that followed after 2 h at 22° C reveals the importance of virus multiplication to local lesion development. *It was the virus replication, movement to, and infection of healthy cells during the 3-4 days at 32° C that respectively elicited and demonstrated HR once the temperature was returned to 22° C.* The experiments of Ohashi and Shimomura (1982) and Konnate and Fritig (1984), in which the sites of virus replication were precisely associated with necrotic lesion formation, suggest that rapid virus multiplication is causally related to host cell leakage and is the basis for necrosis in *N. glutinosa* and *N. tabacum* cv. Samsun NN.

The fact that lesion size limitation and virus replication are associated with the development of necrosis prompted Weststeijn (1981) to examine the temperature control mechanism involved. She measured lesions caused by TMV in *N. tabacum* cv. Xanthi nc following inoculation and subsequent incubation at 22° C for twelve days. Lesion size increased linearly for seven days and peaked at 2.7 mm after eight days. In order to monitor virus spread plants were similarly inoculated, kept at 32° C for 24, 48 and 72 h and then transferred to 22° C for 24 h. Lesion size under this regime increased to 2.0, 4.2 and 7.5 mm, respectively. Clearly, replication and spread of the virus increased with increasing incubation interval at 32° C.

Precise measurements of actual cell damage in the lesion area revealed that the resistance to spread that developed at 22° in cells neighboring the lesion was not totally dissipated at 32° C. It was shown that near complete resistance (suppression of both replication and spread fostered by a 32° C exposure) required at least a 12-day incubation at 22° C prior to the higher temperature treatment. Hence it was postulated that a necrosis inducing factor is synthesized during virus replication and that it is functional at 22° C but not at 32° C where it is gradually degraded. The necrosis factor reported by Hooley and McCarthy (1980) was cited as a possible candidate.

The influence of a toxicity factor in HR development

A necrogenic or toxicity factor, capable of killing tobacco protoplasts, associated with TMV local lesion development was reported by Hooley and McCarthy (1980). The factor was extracted from fully expanded *N. tabacum* cv. Xanthi nc and not from systemic host *N. tabacum* cv. Xanthi. The extract was more toxic to protoplasts from hypersensitive Xanthi nc (42%) than systemic Xanthi (21%). In addition protoplast death was comparatively slow, commencing 6-8 h after exposure to the toxic component and continued for an additional 14 h when cell killing was completed. This time frame was similar to that noted with TMV infections *per se*. The toxin was destroyed when heated at 80° C, but not 60° C, for 5 min. Activity apparently resided in a fraction that eluted from a Sephadex 100 column and coincided with a MW of 4000-8500 and hence seems not to be related to either PR (pathogenesis-related) (van Loon, 1975) or AVF (antiviral factor) (Loebenstein, 1972) proteins.

The site of action of the factor was not established by Hooley and McCarthy. However, its toxicity was exacerbated by 10 mM EDTA and 30-40 mM $CaCl_2$ suppressed protoplast toxicity 80-100%. Since other divalent cations such as Mn^{2+}, Mg^{2+} and Zn^{2+} at 30-40 mM did not affect protoplast survival it would appear that the toxic factor influenced the integrity of the plasmalemma. It is becoming ever more apparent that Ca^{2+} plays an extremely important role in maintaining the selective permeability of the plasmalemma and other membranes (Leshem, 1987). Indeed, cytopathic effects of the factor viewed light microscopically resemble those of dying cells in a necrotic lesion observed electronmicroscopically as reported by Ragetli, 1967.

In an analogous investigation by Modderman *et al.* (1985) evidence was presented that tobacco leaf cell wall fractions reduced mean diameter of TMV lesions. Cell walls were isolated from leaves of 7-week-old Xanthi nc tobacco plants. Leaf cells were macerated with an enzyme mixture from *Trichoderma viride,* Cellulase Onozuka R-10 (Yakult Honsha, Nishinomiya, Japan). A pure carbohydrate fraction of approximately 10 μg/ml reduced lesion diameter by 50% when infiltrated into intercellular space of Xanthi nc tobacco leaves 2 days prior to inoculation with an HR-inducing strain of TMV. A rationale for this phenomenon was not offered; however, perhaps the carbohydrate potentiated HR in a manner that mimicked the aging process in sliced tuber tissue and leaf disks (Tomiyama, 1960) or reflects a cross protection phenomenon (Goodman, 1967).

Graft transmission of HR

Migration of TMV and CMV in respective hypersensitive

and systemic hosts was monitored by Weintraub, Kemp and Ragetli (1961) by grafting normally hypersensitive clones onto systemically infected ones. Such grafts resulted in the development of systemic symptoms in HR hosts. It was contended that systemic invasion results when high levels of inoculum are applied to HR hosts for extended periods of time. It was suggested that in those instances in which mechanical (leaf-rub) inoculation of an HR host results in localization, the site of replication reflects a limited supply of initial inoculum which is too small to overcome the limiting or localizing "barriers". Weintraub and Ragetli (1964a). However, when the graft union permits continuous movement of virus from a larger reservoir of virus, HR is overcome. This, according to Weintraub *et al.* (1960) is similar to what happens when replication in an HR host cell accelerates following a temperature increase from 22° C to 32° C.

Ultrastructural effects of HR

Another perception of the sequence of events leading to TMV-induced host cell death is provided by the electron microscope observations of Weintraub and Ragetli (1964b). They recorded the ultrastructural changes from time of inoculation to cell collapse (over a 78 h time course) in *N. glutinosa* "inoculated with a purified solution of tobacco mosaic virus".

The initial causally related morphological change became apparent 8 h after inoculation as swollen starch grains embedded in an opalescent matrix in chloroplasts (Fig. 29). After 15 h, these starch grains continued to swell suggesting an altered glycolysis. Then at 24 h mitochondrial numbers increased perceptibly and some cytoplasmic membrane disruption was detected. These morphological changes support observed increases in respiration and probably herald changes in TCA cycle activity and a loss of some osmotic control. However, the early starch accumulation perhaps signalled a change in the respiratory pathway suggesting, as we shall see, an NADPH-dependent cytochrome oxidase system liberation of superoxide radicals and oxidization of membrane lipid (Doke and Ohashi, 1988). Degeneration of chloroplasts and mitochondria was evident at 40 h after inoculation when plasmalemma retracted from the cell wall. This situation intensified and 78 h after inoculation cell collapse was observed.

Of additional interest in this study was the failure of the researchers to detect virus particles in the degenerative leaf cells. However, if the number of epidermal and mesophyll cells that were infected initially are so few as 1/50,000 (Sulzinski and Zaitlin, 1982; Steere, 1955), then at least in the early hours after inoculation, evidence of replication and translocation would have been minimal. Nevertheless, if we accept the observations of Otsuki *et al.* (1972) that increases in TMV titer are discernable at 10 h after inoculation, replication must have started sometime earlier. Then an 8 h interval following inoculation for initial replication actually coincides with the ultrastructural change showing apparent altered carbohydrate metabolism noted by Weintraub and Ragetli (1964c) (see Fig. 29).

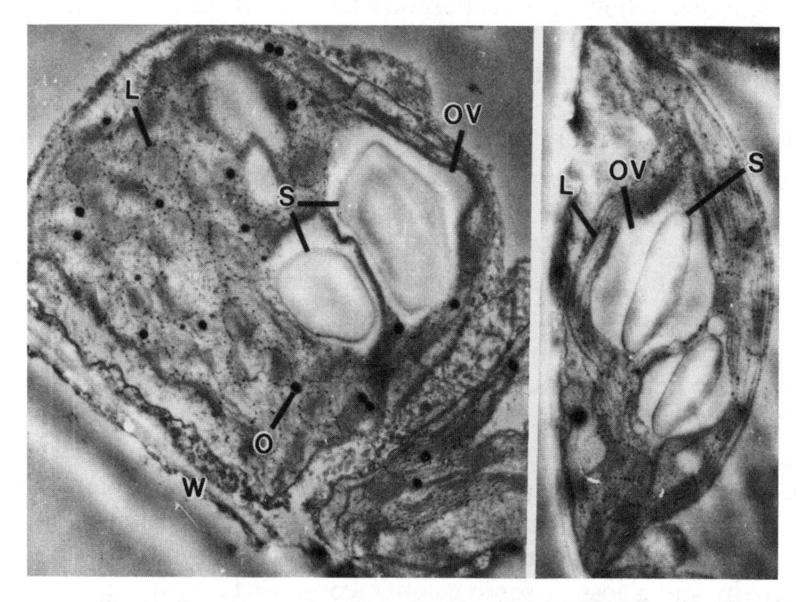

Fig. 29. Cross-section of a chloroplast 15 hours after TMV inoculation. Most of the lamellae are seen in cross-section as a series of points in a circle (*L*). The starch grains (*S*) are greatly swollen and each one is surrounded by an opalescent vacuole (*OV*). Osmiophillic granules (*O*) are dense. Cytoplasmic contents are, at this time, still close to the cell wall (*W*). x 12,000 (left). Chloroplast in cell 8 hours after TMV inoculation. Starch grains (*S*) and the surrounding opalescent vacuoles (*OV*) are greatly swollen, resulting in marked distortion of the lamella (*L*). x 14,500 (right). Weintraub and Ragetli, 1964c.

In a subsequent study, Weintraub, Ragetli and John (1967) were able to detect TMV in small numbers and in minute packets in necrotic *N. glutinosa* leaf cells 40 h after inoculation. Successful virus particle imaging in this series of experiments was attributed to the use of improved fixation and staining with glutaraldehyde and osmium tetroxide. Virus titers could be increased if the inoculated leaves were detached and incubated at 26° C.

It seems that the conditions which control the genetic expression of HR limit the number of host cells that support virus replication. Additionally, it would appear from the observations of Weintraub *et al.* (1967) that the presence of virus in cells portends necrosis and that under conditions favoring systemic infection, even neighboring non-infected cells become damaged and eventually necrose.

Weintraub *et al.* (1967) observed clear differences in virus titer (Tab. 3) and the number of *N. glutinosa* cells that become infected following inoculation with TMV caused either by leaf detachment or raising the incubation temperature after inoculation from 21° to 26° C. These two conditions apparently switched infection from an HR local lesion to a systemic-type. However, crucial to the development of HR resistance was the synthesis of virus in fewer cells and hence fewer necrotic cells.

Tab. 3. TMV Concentration, microgram per milliliter of *N. glutinosa* leaf extract*

After inoculation (h)	Attached leaves	Detached leaves
0	0	0
12	0	0
24	0	11.2
36	0	179.2
40	2.8	-
48	11.2	358.4
60	11.2	358.4
72	0+	179.2

* The figures given are calculated from the highest dilution of the antigen, *i.e.* the leaf extract, that produced an identifiable precipitation line in the agar-gel after immuno-osmophoresis. This dilution end point was equivalent to 0.28 μg purified TMV.
\+ Failure to detect virus serologically at 72 hr is probably a result of too low a concentration in the extract, although TMV can be seen at this time with the electron microscope. Weintraub *et al.*, 1967.

For example, in attached leaves held at 21° C 5% of the cells revealed virus particles at 40 h and the packets of virus were small. However in detached leaves at 21° C 50% of the cells contained virus. When the temperature was raised to 26° C in TMV-inoculated attached leaves, many cells revealed necrosis with small packets of virus and numbers of virus-free cells also showed pathological damage. Of additional interest was the 30 x increase in TMV titer when inoculated *N. glutinosa* leaves were detached and incubated at 26°. Apparently under conditions that foster systemic infection even uninfected cells were damaged and after 72 h those also became necrotic.

Da Graca and Martin (1975) examined ultrastructural changes in TMV-induced local lesions in *N. tabacum* cv. Samsun NN. Their observations revealed virus particles in 5% of the cells they examined only at 72 h after inoculation. Although lesions were apparent at 40 h after inoculation when some cells had collapsed, virus was not detected. As observed by Weintraub *et al.* (1967), it was difficult to reconcile the inability to detect virus particles at 40 h with the degree of cell collapse that was apparent. It is possible that the numbers of particles were too few, 300/necrotic cell according to Weintraub *et al.* (1967), and perhaps they had not yet aggregated into electron dense packets when observed at 40 h.

The sequence of ultrastructural changes that they observed were, however, similar to those reported by Weintraub, Ragetli and John (1967), in that the first ultrastructural change, chloroplast swelling and lamellar distortion noted at 40 h after inoculation, was followed by tonoplast rupture. Mitochondria and nuclei appeared to maintain their structural integrity for 72 h, the time during which cells were collapsing, an observation also made by Weintraub *et al.* (1967).

There was great variability between adjacent cells in the degree of disintegration that was apparent at 40 and 72 h after inoculation in the Da Graca and Martin study. Variations in time of development of host cell disintegration by the different researchers may be explained by the lack of uniformity of inoculum concentration, virus strains, differing local lesion hosts and environmental conditions.

Host cell wall compositional changes in local lesion

Weintraub and Ragetli (1961) recorded cell wall composition alterations as a consequence of local lesion development. The walls of leaf cells surrounding well developed lesions in a radius of 50 cells are largely calcium pectate-containing. In healthy tissues of a comparable age the walls are primarily pectic acid with some calcium pectate. Apparently HR fosters the exosmosis of Ca^{2+} which appeared to be deposited in the walls of a narrow band of cells around the lesion within which virus is not readily detected.

Weintraub and Ragetli alluded to a possible relationship of increased wall rigidity imparted by Ca^{2+} to the hastening of senescence. In this regard Leshem (1981) has indicated that turnover of membrane phospholipid in senescense-associated processes is linked to transmembrane Ca^{2+} flux. A more comprehensive description of the influence of this cation on membrane integrity, as it is associated with active oxygen radical generation, is presented on page 69. Mention is made here to provide a rationale for cytosolic Ca^{2+} flux against a concentration gradient into cell wall free space where it could cross-link pectic acid. Stimulation of active transport by Ca^{2+}-ATPases located on the plasmalemma during senescence has been documented by Kubowitz *et al.* (1982). Efflux of Ca^{2+} may be due to an impaired equilibrium between Ca^{2+} and calmodulin (Leshem, 1987).

Respiratory changes associated with HR

In virus/host combinations in which the plant responds hypersensitively and necrosis develops rapidly (36 h), the rate of respiration is appreciably greater than when infection is systemic (K.R. Wood in Goodman *et al.*, 1986). In an early study by Weintraub, Kemp and Ragettli (1960), a significant rise in respiration occurred several hours prior to the appearance of symptoms (Tab. 4 and Fig. 30). This suggested that the rise was not directly linked to necrotization. A contrary view was reported by Yamaguchi and Hirai (1959), who concluded that the rise in O_2 consumption is directly related to necrotization because necrosis was coincident with the observed increased respiration in TMV- inoculated tobacco.

Tab. 4. Respiration (μl O_2/hour per 100 mg F.W. and per 10 mg D.W.) of leaf disks from *N. glutinosa*, inoculated with TMV (D), and healthy (H).

Hours after inoculation	D		H		Ratio D/H	
	F.W.	D.W.	F.W.	D.W.	F.W.	D.W.
0	30.7	37.0	31.2	38.0	0.98	0.97
6	25.7	29.4	29.1	29.3	0.89	1.00
12	38.0	38.4	35.9	37.7	1.06	1.02
18	37.0	40.8	34.7	41.5	1.04	0.98
*24**	*33.4*	*35.8*	*22.7*	*28.3*	*1.47*	1.27
30	48.4	49.2	26.5	32.2	1.83	1.53
36	41.3	30.0	34.3	39.7	1.20	0.75
42	32.5	13.3	19.2	23.7	1.70	0.56
48	41.1	15.2	26.7	32.5	1.54	0.47

* Italicized data were from the first set of disks in the time series showing the presence of lesions. Weintraub *et al.*, 1960.

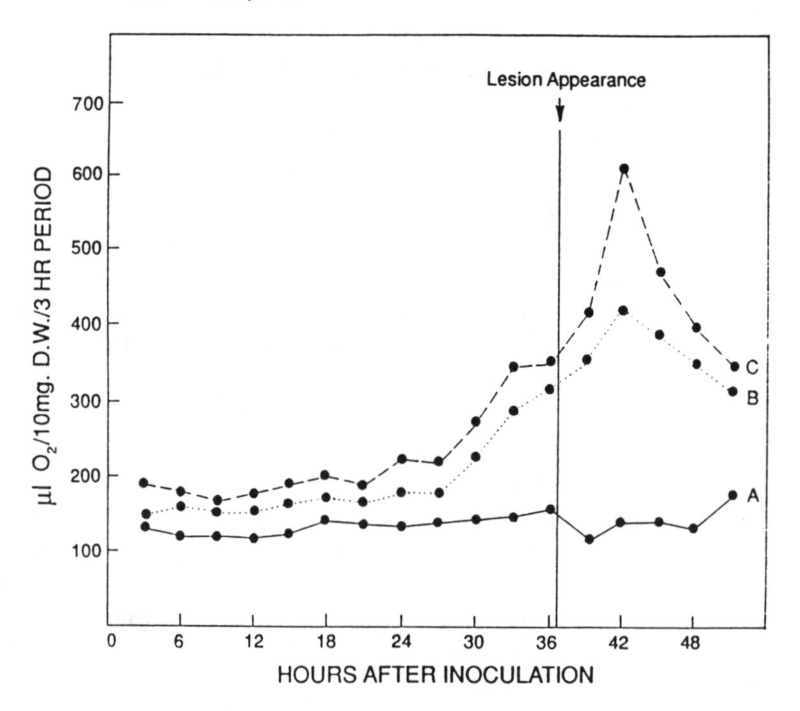

Fig. 30 Respiration as measured by O_2 uptake of disks of healthy and TMV-infected *N. glutinosa* leaves, measured continuously for 51 hours. A: healthy disks; B, and C: diseased disks. Vertical line marked with arrow indicates the first appearance of lesions, 37 hours after inoculation. Weintraub *et. al.* 1960.

Studies by Farkas, Király and Solymosy (1960) and later by Parish, Zaitlin and Siegal (1965) examined some biochemical aspects of the necrotizing phenomenon. Farkas *et al.* measured polyphenoloxidase (PPO) activity in *N. glutinosa* inoculated with the "para" strain of TMV and sampled 2-3 days *after* inoculation, "approximately one day after the appearance of lesions". They reported no consistent increase in PPO prior to the appearance of symptoms, "although a slight trend" towards activation of the enzyme precedes lesion formation. That lesion development was severely reduced by providing reducing power in the form of ascorbic acid was also apparent. Similarly, Parish *et al.* reduced lesion darkening in *N. sylvestris* inoculated with the U_2 strain of TMV, by exposing the tissue to ascorbic acid and intensified necrosis (tissue darkening) by supplying additional substrate such as cafeic acid, chlorogenic acid and others.

Unfortunately cells treated with either reducing agents, or substrate for potentiated quinone accumulation were not examined ultrastructurally. It would have been of great interest to determine whether cell membrane integrity was influenced by either of these treatments. In retrospect, and from the recent data concerning oxygen radical disorientation of membranes, it is possible to speculate that these treatments would have affected both tissue darkening and membrane integrity *per se*. This contention is indirectly supported by experiments conducted by Parish *et al.*, who supplied phenols (catechol, chlorogenic acid) to tissues inoculated with the *systemic* U_1 strain, and failed to produce necrosis. They concluded sagely that more is involved in HR-related necrosis than PPO and accumulation of its substrate.

It is clear however from the data of Parish *et al.* that when *N. glutinosa* is inoculated with either the U_1 (systemic) or U_2 (necrotic) strain of TMV, increased O_2 consumption is quickly induced by the necrotic or HR inducing form. Most importantly, *the level of virus replication induced by either strain is essentially the same for at least 30 h* (Fritig *et al.* 1987). Hence, the increased O_2 consumption cannot be linked directly to virus replication. However, the study by Weintraub *et al* (1960) attributed the rise in O_2 consumption to virus "activity". Clearly virus replication is implicit in HR development, and an increase in O_2 consumption and necrosis follow.

Dwurazana and Weintraub (1969a) followed the respiratory patterns of *N. tabacum* in response to infection by four strains of potato virus X (PVX). They noted that the rise in respiration occurring prior to the appearance of symptoms was correlated with the severity of symptoms. Using specific enzyme inhibitors, it appeared that neither glycolysis nor oxidative phosphorylation (Krebs cycle) was altered. However the inhibitor sodium fluoride (NaF) did not have marked effects on either healthy or infected tissue. Since glycolysis is inhibited by NaF and the pentose phosphate pathway is not, it was suggested that the *normal-appearing respiration* in infected tissues *reflected pentose phosphate pathway activity*. In examining that possibility carefully, Dwurazana and Weintraub (1969b) observed that indeed the pentose pathway was extremely active in tissues infected by all 4 strains of PVX (Tab. 5). In addition, all strains decreased the C_6/C_1 ratio of healthy and PVX-infected tobacco leaves (Fig. 31).

Tab. 5. Activity of enzymes of the pentose phosphate pathway in cell-free extracts of healthy, and of PVX-infected tobacco leaves with well-developed symptoms

	Enzymic activity in µmoles/minute/gram fresh weight				
		Potato virus X strains			
Enzyme	Healthy	Yellow	Mild mottle	Brown spot	Ring-spot
Glucose-6-phosphate dehydrogenase	0.09	0.125 (1.39)*	0.131 (1.46)	0.156 (1.73)	0.170 (1.89)
6-Phosphogluconate dehydrogenase	0.064	0.087 (1.36)	0.101 (1.58)	0.121 (1.89)	0.130 (2.03)
Phosphoriboisomerase	0.056	0.052 (0.93)	0.031 (0.55)	0.040 (0.71)	0.038 (0.68)

*The figures in parentheses express the ratio of infected: healthy for each strain and for each enzyme. Dwurazana and Weintraub, 1969a.

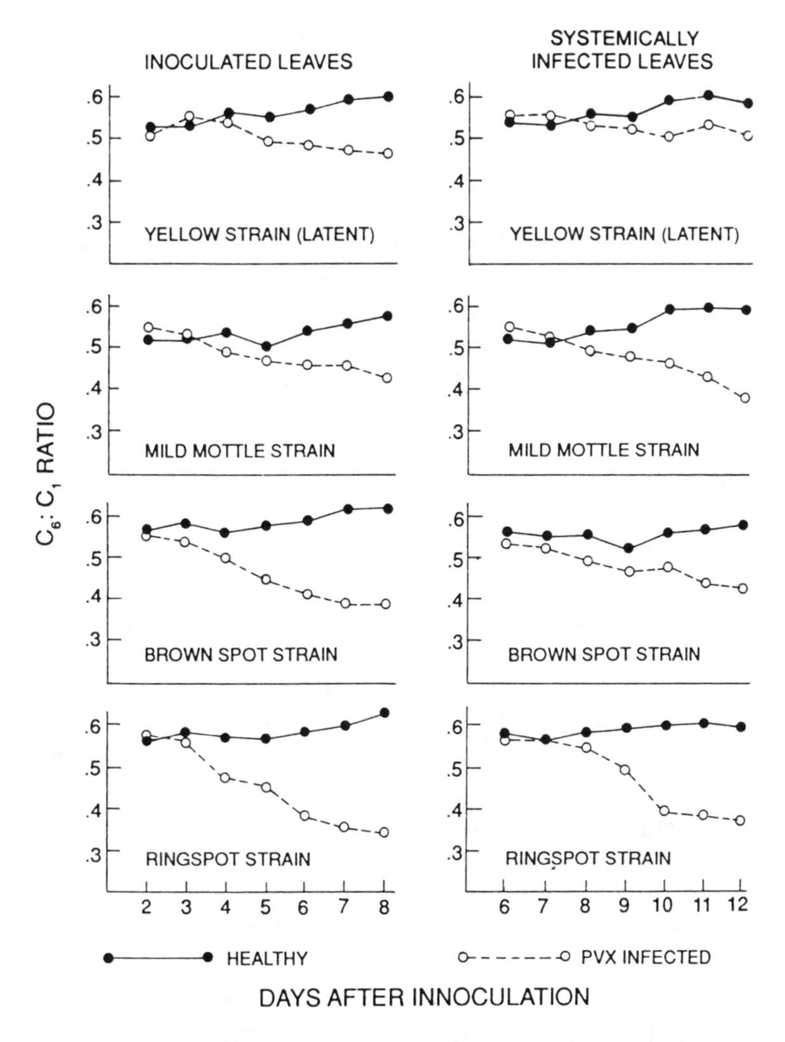

Fig. 31. Changes in the C_6/C_1 ratio of healthy and PVX-infected tobacco leaves. Dwurazana and Weintraub, 1969b.

There is of course some question as to whether the apparent stimulation of the pentose phosphate pathway in tobacco by the ringspot strain of PVX also occurs in tobacco inoculated with a local lesion-inducing strain of TMV. It must be pointed out that clear evidence for PVX-activated pentose phosphate enzymes was detected in tissues already showing well developed symptoms. Similarly, Farkas *et al.* (1960) noted a comprehensive shift to the pentose phosphate pathway in *N. tabacum* cv. White Burley inoculated with the "para" (HR^+) strain of TMV. However, their analyses for specific enzyme potentiation were made from samples 4 - 6 days after inoculation. It is apparent that at that time HR had long since been triggered and the basic metabolic perturbations that cause membrane damage had already occurred.

The question may be raised here whether other respiratory enzyme systems, for example NADPH oxidase, are responsible for the occasionally observed pre-symptom O_2 consumption increase. It is indeed likely that the precipitous rise in respiration noted *after* symptoms have developed is due to PPO activity on substrate released due to host cell membrane destruction.

Relationship of oxygen radicals to virus-induced local lesion formation.

It has generally been held, without definitive correlative evidence, that HR in potato induced by *P. infestans* and manifested as rapid cell death and necrosis was the same as that induced by *Pseudomonas syringae* pv. *pisi* in tobacco (Samsun NN) and by TMV in the same host. The hypersensitive reaction (HR) has been described by Pryor (1987) as the predominant mode of expression of single major gene resistance in plants against *fungal* pathogens. This description is probably as fitting for the interaction between plants and some *viral* pathogens.

Plants have developed a particularly efficient and fundamental defense mechanism in the hypersensitive reaction. However, they are not unique in this regard as the animal kingdom is capable of generating the basic reactions integral to HR. Though plants are not protected against disease and assorted trauma-inducing insults by antibodies and macrophages, in both the plant and animal kingdoms the generation of oxygen-centered free radicals has been documented. Ground state O_2 or "dioxygen" may be altered to the triplet molecule which may be successively reduced to $O_2^{\cdot-}$ (superoxide), $OH^{\cdot}$ (hydroxyl radical), $^1O_2$ (singlet oxygen), H_2O_2 (hydrogen peroxide) and others by biochemical, chemical and photodynamic reactions. These highly reactive molecules are destructive particularly to the cell's membrane systems and presents the cell with serious problems for survival. In the course of evolution, plants have developed safeguards to control the levels of these reactive molecules; catalase, peroxidase, polyphenol oxidase and specifically superoxide dismutase are examples of the concerned regulatory enzymes.

The precise oxygen radical(s) that injures plant cell membranes during the development of HR has as yet not been determined. However, the direct influence of the oxygen radical on plant cell membranes during the course of the hypersensitive reaction has, as we shall see, been reported. Indeed, there is good evidence that oxygen radicals are generated in HR as it develops in response to viral, bacterial and fungal pathogens.

Most of the experiments suggesting the participation of oxygen radicals in the development of HR rely on suppression of the reaction by superoxide dismutase (SOD) and catalase. Hence, Elstner (1982) suggested that ·OH is the active species. The short "life time" of free ·OH indicates that its activity is generally restricted to the site of its generation. The highly localized necrosis that develops in HR-responding tissues clearly obviates the necessity of extensive migration of this oxygen radical.

The probability that singlet oxygen may participate in HR would require evidence of the participation of photodynamic processes. Indeed, the observation that dark treatments of virus-infected tissues diminishes HR intensity would suggest that singlet oxygen could also be a candidate.

That organic peroxides or peroxy radicals (ROO•) may be the cause of HR is difficult to determine bacuse they share similarity in substrates with ·OH, for example. Both are able to oxidize unsaturated fatty acids to form malondialdehyde (Kato and Misawa, 1976).

The relationship of free radical formation to lignification, which requires the synthesis of phenylpropanoid precursors, peroxidase and NADPH, is probably derived from wall-bound malate dehydrogenase (Peng and Kúc, 1992). It also seems possible, although not indicated by Gross *et al.*, (1977), that the required NADPH could be the product of potentiated hexose monophosphate shunt activity. Subsequent shikimic acid pathway and tyrosine/phenylalanine synthesis of the lignin precursors of monophenols that accept electrons from oxygen radicals may contribute to the later developing necrogenesis in tissue undergoing HR.

Perhaps the earliest report of the involvement of highly reactive oxygen radicals in the development of HR cell death and necrosis as a consequence of a plant virus infection was by Kato and Misawa (1976). In an even earlier brief report they clearly discounted the long held notion that the oxidation of phenols was responsible for HR-reacting CMV-infected cowpea leaves. Rather, they contended that HR was brought about by oxidative processes under the control of lipoxygenase. This enzyme is known to act on a number of polyunsaturated fatty acids whose oxidative deterioration products include free radical intermediates such as $O_2^{\cdot -}$, singlet oxygen, $OH^{\cdot}$ hydroxyl, $RO^{\cdot}$ alkoxyl, $ROO^{\cdot}$ peroxyl, semiquinone free radicals, H_2O_2 and others.

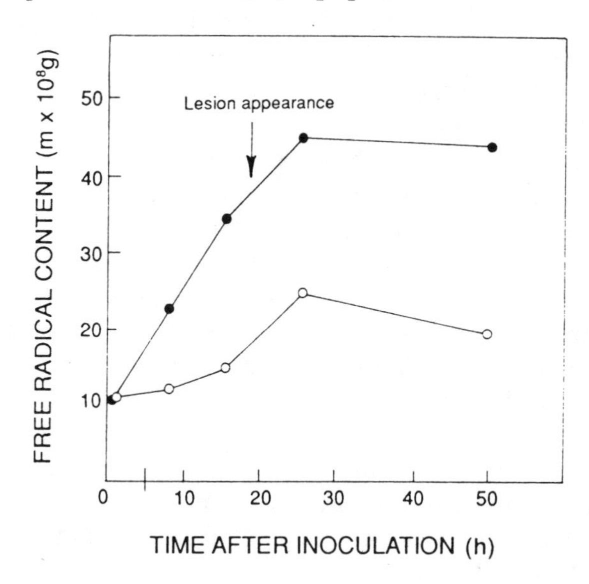

Fig. 32. Changes of free radical content in CMV-infected cowpea leaves. These values are based on comparison of the absorption given by approximately 0.6g samples with the absorption given by 10^{-8} M of diphenyl picryl hydrosol free radical under identical instrumental conditions O: healthy leaves ●: infected leaves. Kato and Misawa, 1976.

In their second report, presented in some detail below, Kato and Misawa (1976) described the formation of free radicals (Fig. 32) and the accumulation of malondialdehyde (MDA) (Fig. 33), the decomposition product of lipid peroxides, as well as changes in polyunsaturated fatty acid content (PUFA) which causes cell membrane permeability alterations during the development of HR in CMV-infected cowpea leaves.

Malondialdehyde increased markedly during the 8 - 15 h after inoculation and peaked just prior to lesion appearance. Evidence that the sharp increase of malondialdehyde reflected lipoxygenase activity was gained by suppressing the compound's accumulation with a specific inhibitor of lipoxygenase, dibutylhydroxytoluene (BHT). When the inhibitor was applied to the CMV-inoculated tissue, not only was malondialdehyde accumulation suppressed, but in addition, no visible symptoms appeared (Fig. 33). The target molecules in the cellular membranes, the polyunsaturated fatty acids linoleic and linolenic acids, decreased to 18.5 and 22.0%, respectively, of those in uninfected leaves during the 24 h course of the experiments.

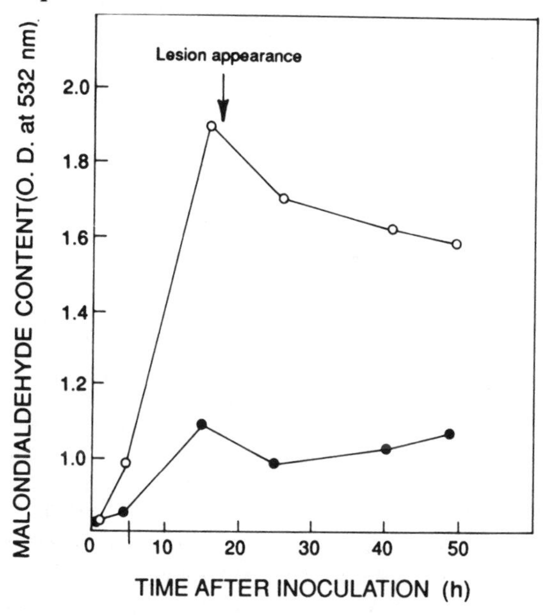

Fig. 33 Effect of dibutylhydroxytoluene on malondialdehyde production in CMV-infected cowpea leaves. Malondialdehyde contents were expressed as rate of control. ○ :untreated ● :treated. Kato and Misawa, 1976.

If HR occurs as a result of a chain reaction producing free radicals by the peroxidation of unsaturated fatty acids, Kato and Misawa reasoned that it should be possible to prevent necrosis in CMV-infected tissue with free radical scavengers. The three oxygen radical scavengers applied, 2 mercaptoethylamine (MEA), 2-aminoethylisothiouronium bromide hydrobromide (AET) and 1, 1 diphenyl-2-picryl hydrazyl (DPPH) all inhibited lesion formation nearly 100% at 0.01 M concentration. Suppression of malondialdehyde by DPPH is shown in Fig. 33.

Electrolyte leakage from infected tissue increased nearly 12x over healthy leaves in 6 h. The time course of electrolyte flux is shown in Fig. 34 (Kato and Misawa, 1976).

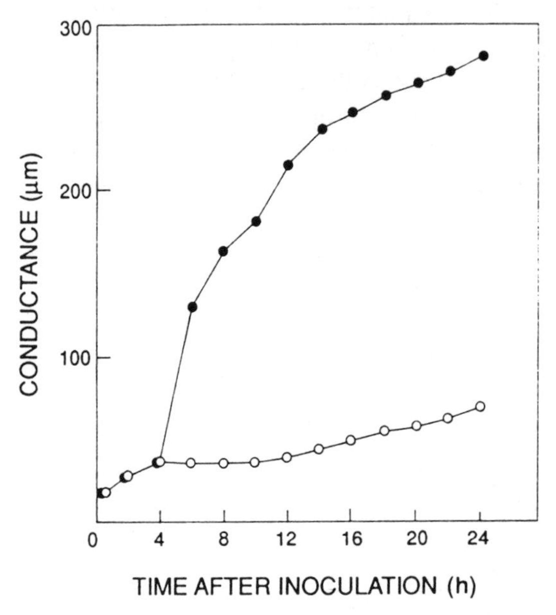

Fig. 34. Effect of infection on loss of electrolytes from cowpea leaves O : healthy leaves ● : infected leaves. Kato and Misawa, 1976.

It appears from these experiments that as a consequence of infection of cowpea by the local lesion-causing yellow strain of CMV, comprehensive amounts of free radicals are released. They are released over a time course that accounts for lesion development and electrolyte flux. *Lipoxygenase activity* was established and the near 80% decrease in PUFA *appears* to *account for the free radicals* released and the necrosis that develops.

Changes in free radical content of cowpea leaves infected with CMV were monitored by paramagnetic resonance adsorption. Free radical levels twice that of the healthy control were detected 3 h after inoculation which coincided well in time with the membrane leakage detected in *Gomphrena globosa* caused by TBSV (Pennanzio and Sapetti, 1982). *The level of free radicals more than tripled prior to appearance of visual symptoms.*

HR-related changes in fatty acids and phospholipids of plant cell membranes caused by TMV were also studied by Ruzicska, Gombos and Farkas (1983), perhaps one of the last of many major contributions from the laboratory of Gábor Farkas in Szeged, Hungary. The impetus for their study was the astute recognition that HR causes an abrupt cell permeability change that is preceded by membrane autolysis. As a consequence, changes in membrane lipid composition were studied in tobacco following temperature shifts from 33° to 25° C after inoculation with TMV.

Analysis of lipids from the microsomal fraction (exclusive of chloroplasts and mitochondria) revealed changes in the proportions, eg. palmitic acid increased 10% while linolenic acid decreased prior to the appearance of symptoms. Fig. 35 indicates that prior to symptoms, all double bond-containing fatty acids were reduced in number while the saturated fatty acids, particularly palmitic acid rose dramatically.

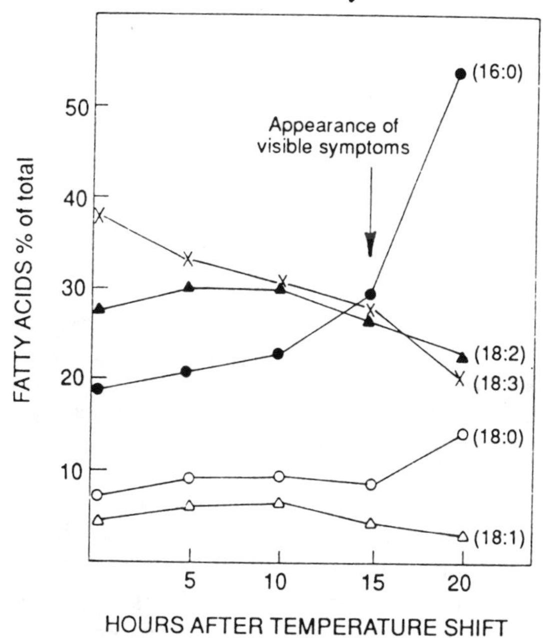

Fig. 35 Changes in the fatty acid composition of microsomal membrane fractions from TMV-infected *N. tabacum* cv. Xanthi nc after the temperature shift from 33° to 25°. Ruzicska *et al.*, 1983.

In Fig. 36 increased *lipoxygenase activity* which in turn supported a dramatic increase in symptoms appeared to coincide with these fatty acid changes. These data show an exceedingly sharp increase in electrolyte leakage coinciding with the time of symptom appearance. This apparent change in membrane permeability appeared to have its genesis about 2 h prior to the appearance of symtoms.

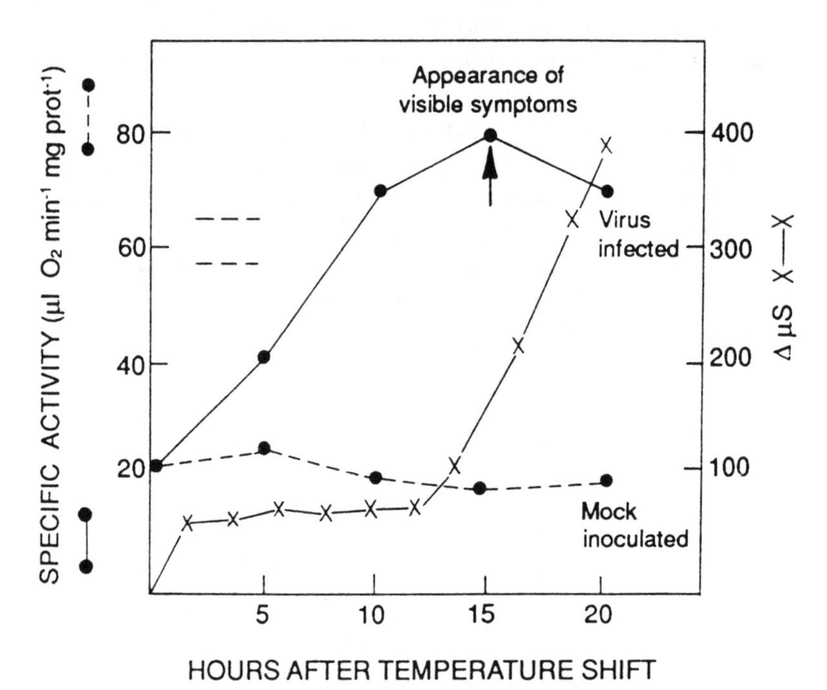

Fig. 36. Lipoxygenase activity in TMV-infected (●—●) and mock inoculated (●---●) *N. tabacum* cv. Xanthi nc and the leakage of electrolytes in infected (X—X) *N. tabacum* cv. Xanthi nc following transfer of the plants from 33° to 25°. Ruzicska *et al.*, 1983

One could conjecture from these data that membrane lipids slowly oxidized from 8-10 h after shifting the TMV inoculated plants from 33 to 25° C. Once a threshold level of membrane disorientation is reached, its dysfunction precludes continued osmotic selectivity and perhaps endogenous scavenger molecules could no longer protect the membranes. Hence, leakage was intense and cell death and necrosis followed swiftly. As we have already noted, similar results were published by Kato and Misawa (1976).

The actual lipid change that develops is a physico-chemical one in which the fluidity of the unsaturated lipids is altered to a more rigid hydrogenated state. According to Leshem (1987), phospholipase A_2 then proceeds to release linolenic acid and linoleic acid from the membrane. These are substrate for lipoxygenase which in turn generates oxygen free radicals. Hypothetically then, when the free radical level rises to a certain point, membrane dysfunction occurs abruptly.

The relationship of lipid peroxidation to HR-induced necrogenesis and tissue collapse was studied by Beleid El-Moshaty *et al.* (1993) (Fig. 37) in cowpea (*Vigna unguiculata)* infected by either tobacco ringspot virus (TRSV) or southern bean mosaic virus (SBMV). TRSV causes HR local lesion symptoms, whereas SBMV, a positive control causes mosaic mottling. They observed lipid peroxidation 12 h prior to any significant K^+ efflux and 9 h before HR-local lesions appeared. Lipid peroxidation, as measured as MDA accumulation, continued to increase during lesion development and declined at 72 h after infection which was coincident with tissue collapse. Elevated $O_2^{\cdot-}$ production, measured as NBT reduction, and SOD activity (the inhibition of NBT reduction) paralleled the increase in lipid peroxidation. *This sequence of events suggested that structural membrane damage, effected by lipid peroxidation, precedes electrolyte leakage and tissue collapse.* The data were also interpreted as indicative of $O_2^{\cdot-}$ generation being directly responsible for the membrane lipid peroxidation. Indirect support for this contention was the observation that lipid peroxidation proceeded without a concomitant increase in lipoxygenase activity. These observations appear to be at odds with those of Ruzicska *et al.* (1983); however, their system was involved with TMV-infected Xanthi nc tobacco. In *addition, these two studies may reflect the operation of a two-step process wherein oxy-radicals initiate membrane lipid peroxidation, which in turn supplies substrate for lipoxygenase activity.*

Doke and Ohashi (1988) have more clearly linked $O_2^{\cdot-}$ generation to HR in tobacco infected with TMV. The generation of $O_2^{\cdot-}$ was measured photometrically as the reduction of exogenously supplied cytochrome *c*. In these experiments the reduction of extracellular cytochrome *c* in Samsun NN or Samsun leaf disks previously inoculated with TMV was determined following

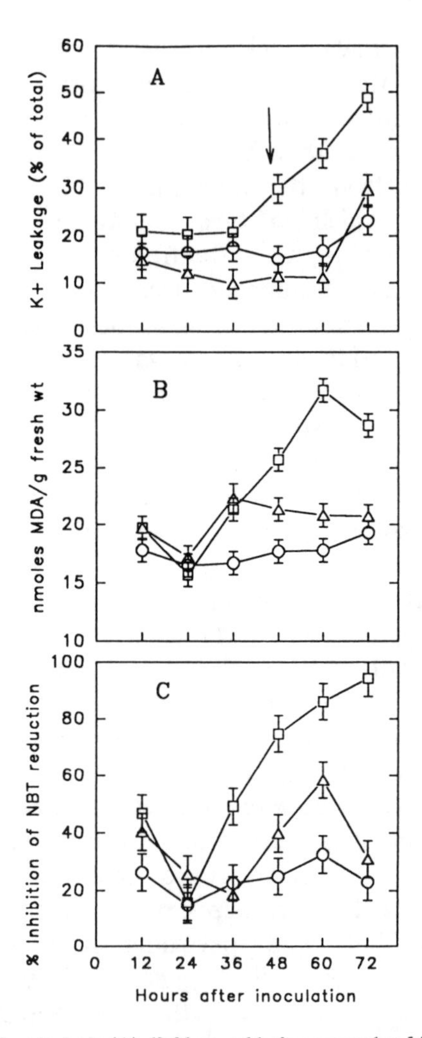

Fig. 37. Potassium leakage (A), lipid peroxidation assayed as MDA content (B) and SOD activity measured as inhibition of NBT photoreduction of formazan (C) in TRSV □, SBMV **Δ**, and buffer-rubbed control (O) cowpea leaf disks. Arrow denotes time of lesion formation in TRSV-infected leaves. Beleid El-Moshaty *et al.*, 1993

their transfer from 30° C to 20° C. Measurements were made immediately after transfer and at 2 min intervals thereafter (Fig. 38). The infected leaf disks, but not the uninfected ones, revealed a decisive cytochrome *c* reducing activity immediately after the transfer to 20° C. This reducing activity could be lowered to negligible amounts when superoxide dismutase (SOD) was added to the reaction mixture.

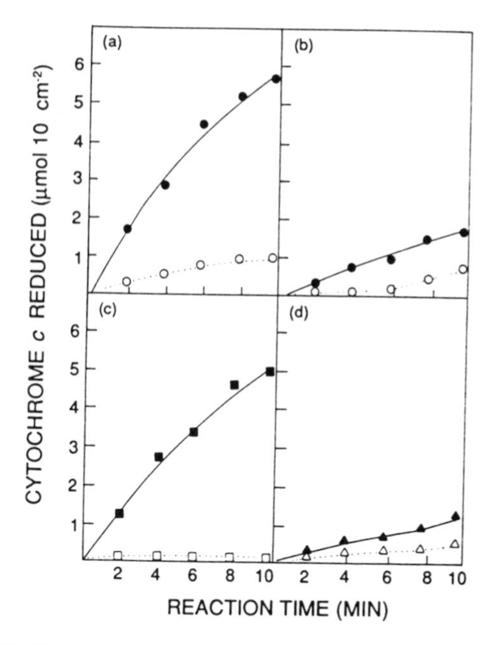

Fig. 38. Extracellular cytochrome *c* reducing activity of tobacco leaf disks. The disks were inoculated with TMV and then incubated at 30° C for 36 h before transfer to 20° C. The controls were treated in the same way except that they were sham-inoculated with buffer. Enzyme activity (NADPH oxidase) was assayed immediately after transfer to 20° C. Four disks (18mm in diameter) were added to a reaction mixture consisting of 20 μM cytochrome *c*, 0.1mM EDTA and 0.01 mM potassium phosphate (pH 7.8) with or without SOD (50 μg/ml) as required. (a) TMV-inoculated tobacco cv. Samsun NN ● reaction mixture without SOD; O reaction mixture with SOD (b) Sham-inoculated cv. Samsun NN ● reaction mixture without SOD, O reaction mixture with SOD; (c) TMV-inoculated cv. Samsun NN ● enzyme assayed in reaction mixture without SOD at 20° C (■) and at 30° C (□); (d) Samsun enzyme assayed following TMV-inoculated (▲) and sham-inoculated (Δ), both without SOD in reaction mixture. Doke and Ohashi, 1988.

The membrane fraction from inoculated Samsun NN also demonstrated cytochrome *c* reducing activity when shifted to 20° C in the presence of both NADPH and $CaCl_2$. Obviously the NADPH oxidase had been activated in both systems, and in both the oxygen radical scavenger SOD was able to bring the reducing activity to very low levels. Additionally, when the experiments were done in vitro and $CaCl_2$ was eliminated from the reaction mixture, reducing activity was partially eliminated. It was totally eliminated when NADPH was not supplied. Thus, the temperature shift from 30° to 20°C required for local lesion development brought with it the activation of NADPH oxidase reduction of cytochrome *c* and the generation of $O_2^{\cdot-}$.

Induction is swift as the reaction is activated immediately on temperature shift and necrosis develops in the subsequent 6-7 h. The net cytochrone *c* reducing activity, detected in the *brief 10 min* following transfer of the disks to 20° C was correlated with the ring shaped lesions that appeared on each disk after incubation at 20° C (Fig. 39).

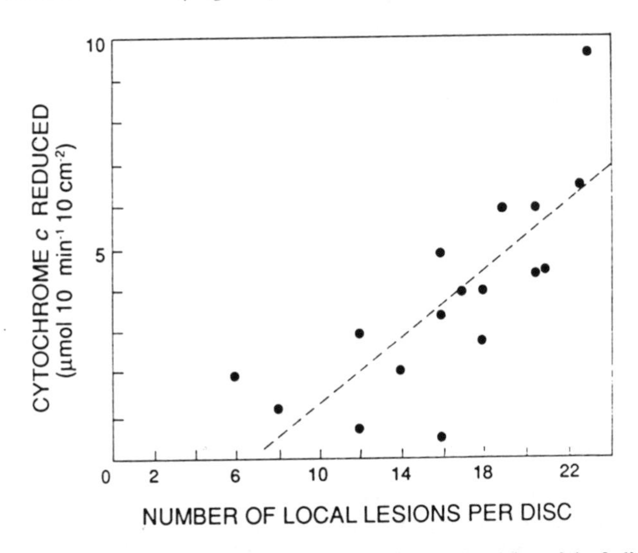

Fig. 39. Relationship between O_2. generating activity of leaf disks of cv. Samsun NN. 10 min after transfer to 20° C and the number of necrotic lesions which appeared 24 h after transfer to 20° C. The experimental conditions were the same as for Fig 38. Each point is the cytochrome c reducing activity in TMV-infected leaf disks minus the activity in non-infected control disks. Regression line $y = 0.4x - 2.7$, $r = 0.75$, $P < 0.01$. Doke and Ohashi, 1988.

It seems that, as in potato infected with *P. infestans*, an $O_2^{\cdot-}$ generating reaction is an early and integral biochemical event in the development of the necrotic local lesion in tobacco species carrying the N gene for resistance to TMV. The reaction was suppressed when $NADP^+$, SOD or catalase were infiltrated into the TMV-infected Samsun NN leaf disks. This response reflects the scavenging of $O_2^{\cdot-}$ by the two enzymes and substrate deficiency when the oxidized nucleotide was supplied. It may be possible now to explain some of the earlier biochemical, microscopical and ultrastructural observations on the basis of these findings. An important question that remains is to what level virus synthesis must rise before symptoms develop. (See Fig. 23, page 78). This gap in our understanding persists despite the advances that have recently been made concerning the genetic control of HR.

Role of the N and N' genes in HR development

The history of the N gene and its genealogy in the genus *Nicotiana*, first described for its necrotic response to TMV by Holmes (1929), was recently retraced by Dunigan, Golemboski and Zaitlin (1987). When Holmes described the response of tobacco to TMV, he also recognized its value as a bioassay for infection. The N gene in *Nicotiana* is a single dominant gene whose expression is temperature sensitive in tobacco species. As described in previous sections, at 20-22° C the gene is activated by TMV infection and necrosis develops rapidly; however, at 28° C and above infection becomes systemic, being expressed as a typical light and dark green foliar mosaic pattern.

Although much has been written about HR as it has been studied in tobacco inoculated with TMV, many of the reports, in our opinion, record phenomena that are not integral to HR. The salient feature of HR, from our point of view, is its induction of rapid host cell death characterized by cell collapse and subsequent necrosis. It is becoming clear that the HR-linked (rapidly developing and restricted) necrogenesis is a consequence of the HR-induced membrane damage, cell disorganization and collapse. In fact if HR develops in total continuous darkness HR proceeds, but necrogenesis does not (Peever and Higgins, 1989).

Despite the comprehensive literature, pertinent in varying degrees to HR, several essential features of the phenomenon remain unknown. Neither the specific nature of the inducer nor the site of its action was until recently established; however, it would appear that the HR in virus-infected plants is indeed an example of the gene for gene response. As we indicated earlier the length of time the two must react with one another is, under defined conditions, known (see Fry and Matthews, 1963 and Weststeijn, 1978, 1981). An inkling of the nature of the inducer (Hooley and McCarthy, 1980; Modderman *et al.*, 1985) has previously been mentioned. As we shall see, some gene constructs (Knorr and Dawson, 1987; Takamatsu *et al.*, 1987) and (Padgett and Beachy, 1993) have located both the N' and N genes as well as a possible site of action on the plasma membrane (Kato and Misawa, 1976; Doke and Ohashi, 1988).

With respect to the isolation of a host gene or genes responsible for inducing HR in N gene tobacco, the road has been an arduous one. The Zaitlin group, Dunigan *et al.* (1987), have reported that in their search for HR-specific genes more than 134,000 cDNA clones were screened without success. In their system the host is *N. tabacum* cv. Xanthi nc and the virus is the U_1 severe strain. What appears to be a breakthrough in this regard has been reported by Padgett and Beachy (1993) and will be discussed at the close of this section.

Recently Knorr and Dawson (1987) used *N. sylvestris*, a species that contains the N' resistance gene, to study HR resistance genetically. The N' gene reacts *systemically* with the common (mild) strain of TMV. However, it was possible to isolate mutants of the mild strain that induce HR in *N. sylvestris.* By constructing genomic recombinants between the common strain and a local lesion-inducing mutant they were able to identify the location of the mutation responsible for the HR phenotype (Fig. 40). This was done by exchanging cDNA fragments between the mutant and its wild type (mild strain of TMV) parent to construct a set of defined viral recombinants. In order to expose the specific sequences conferring the mutant phenotype, the recombinants were inoculated onto leaves of *N. sylvestris*. With this process they identified a point mutation in the coat protein (CP) gene of a TMV mutant that caused HR in *N. sylvestris.* Sequences from the mutant that converted the hybrid genome to mutant phenotype (inducing HR in *N. sylvestris*) contained only a single nucleotide difference from the wild-type parent virus. A cytosine-to-uracil alteration in this construct changed the capsid protein to encode phenylalanine instead of serine. *Hence this transcript, which expressed the mutant phenotype, indicated that the single nucleotide substitution was sufficient to induce local lesions (HR) in N. sylvestris.*

However, an additional point mutation was detected which consisted of a nucleotide change of adenine to guanine, in the 3' nontranslated region. When each of these point mutations was individually substituted into the wild-type background, only the phenylalanine for serine substitution in capsid protein caused HR.

The point mutation at the 3' nontranslated region produced systemic mosaic symptoms when substituted into the wild-type background.

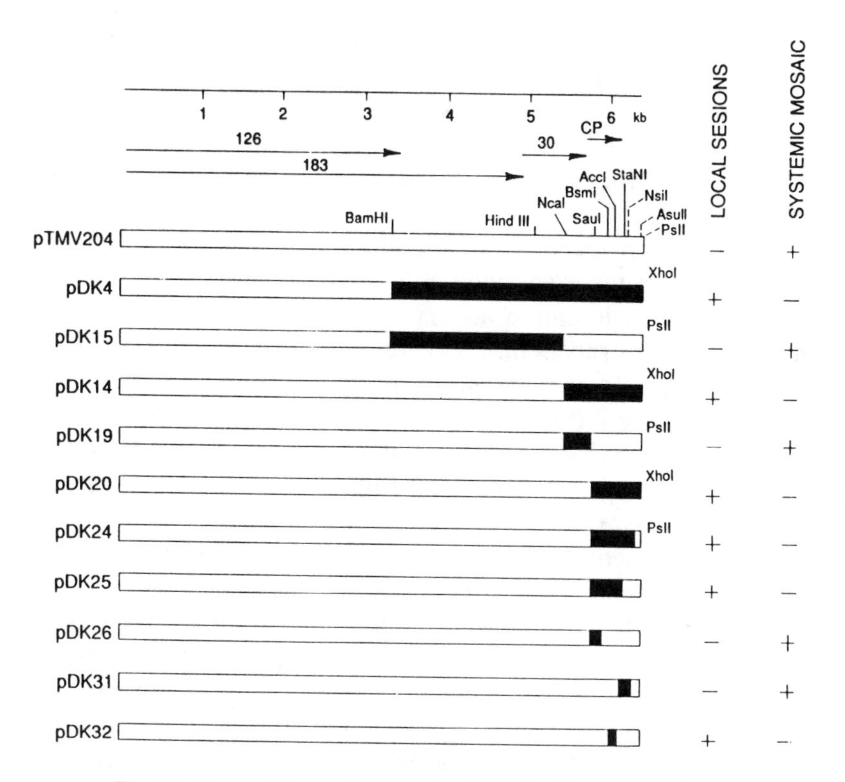

Fig. 40. Composition and symptomatology of wild-type and 10 recombinant TMV genomes. Dark portions of genomes show sequences from TMV204.D1, a mutant that causes local-lesion formation on *N. sylvestris*. Light portions of genomes represent sequences from the wild-type parental strain, TMV204. Relative positions of restriction endonuclease sites used in engineering the constructs are shown for pTMV204. The 3' termini derived from pDK4 contain a *XhoI* restriction site, whereas all other 3' termini contain a *Ps11* restriction site. Symptoms produced on *N. sylvestris* plants inoculated with *in vitro* transcripts of the plasmids are indicated beside each genome. The size of the TMV genome in kb is represented above. Arrows represent open reading frames with the mass of product in kilodaltons; CP, capsid protein. Knorr and Dawson, 1987.

It would appear therefore, at least in N' gene-carrying-*Nicotiana,* that capsid protein, in addition to protecting and stabilizing the virion, has a second function: *that of establishing virus-host compatibility during the infection process.* Analogously, Knorr and Dawson noted that sequence alterations *outside* the capsid protein gene may induce N' gene HR.

In a subsequent publication (Culver and Dawson, 1989a) four additional point mutations in different regions of the CP gene of TMV were defined. Each was produced by site-directed mutagenesis of cDNA and each produced a phenotypically distinct HR on both *N. sylvestris* (N' genotype) and *N. tabacum* (N genotype). The four nucleotide alterations made were guanine to adenine, cytosine to uracil, adenine to guanine and a second adenine to guanine mutation. These resulted, respectively, in amino acid replacements of methionine for valine, leucine for proline, serine for asparagine and glycine for asparagine.

The single amino acid substitution controlling elicitation suggests that stereochemistry of the receptor on the host cell is precise. It would appear, therefore, that tertiary and quaternary structure of the CP is crucial. *The study of Culver and Dawson (1989a) suggests that the more copies of the coat protein monomer present, the stronger the reaction. This certainly implies that a level of replication is a priori necessary for the induction of HR* in the TMV-N' interaction. It was also clear that a number of sites in the CP N' gene of TMV affect host response.

It is of historical interest and fundamental for the purpose of accreditation, that we mention the fact that earlier studies of Funatsu and Fraenkel-Conrat (1964) and Wittmann and Wittmann-Liebold (1966) suggested that CP of TMV probably played a role in inducing HR in *N. sylvestris*. Culver and Dawson (1989a) point out that, although each of their CP mutants induced HR in *N. sylvestris*, not all mutations will do so. In fact the report of Funatsu and Fraenkel-Conrat described CP amino acid substitutions that did not induce HR. In a study by Dawson, Bubrisk and Grantham (1988), it was indicated that TMV is capable of causing HR development in *N. sylvestris* despite large deletions in its CP gene. We must conclude therefore, that a variety of conformational configurations on the TMV virion surface will complement an array of receptors on the tobacco leaf epidermal cell (protoplast) surface. The variations in phenotypic response with respect to lesion size and time of appearance tend to support this contention.

HR induction by altered RNA or protein?

In order to determine whether the altered viral RNA or the altered protein acted directly as an elicitor of HR (Culver and Dawson, 1989b), CP translational starts were removed from the HR-inducing mutant TMV25 and the systemically infecting (mosaic-causing) TMV U_1 strain. The production of mosaic symptoms by both mutants, albeit slowly, and the absence of HR symptoms suggested the necessity of CP for elicitation. This was confirmed by the detection of infectious RNA from both mutants in inoculated and systemic mosaic leaves and the absence of CP from both mutants. This study demonstrated the essentiality of TMV25 CP in the elicitation of HR in *N. sylvestris*.

In addition, infectious RNA was not detectable in leaves that were not directly inoculated prior to the time that mosaic symptoms eventually developed. This finding suggested to Culver and Dawson (1989b) that CP may also influence rapid long distance movement of TMV.

Lindbeck *et al.* (1991) revealed that accumulations of coat protein in transgenic plants, which did not produce local lesions, were many times lower than in infected plants; hence, a threshold of elicitor protein is required to induce HR. Also, in NN protoplasts, a cluster of protoplasts, rather than a single cell is needed for the expression of HR necrosis. It would appear that an amplification of the induction process by several plant cells is required for HR development.

To further examine the role of these CPs in the induction of the N' form of HR, both elicitor and non-elicitor CP open reading frames (ORF) were transformed into the genome of *N. sylvestris* (Culver and Dawson, 1991). Transgenic *N. sylvestris* plants expressing a non-elicitor CP displayed a phenotype indistinguishable from non-transformed healthy control plants. Transgenic *N. sylvestris* plants expressing the elicitor CPs displayed stunting and developed necrotic patches that eventually coalesced and collapsed entire leaves. The severity of necrosis produced by transgenic plants correlated with the type of elicitor CP expressed. Hence, plants that expressed a strong elicitor CP displayed greater severity of necrosis when compared to plants expressing a weak elicitor CP. This display of an HR phenotype in transgenic plants expressing elicitor CPs further demonstrates, according to Culver and Dawson (1991), that these CPs are singly responsible for the induction of the HR in *N. sylvestris.*

Additional support for the elicitor-receptor concept, which is now generally predicted (Gabriel and Roth, 1990; Keen, 1990), has been provided by Culver, Dawson and Stubbs (private communication) in an analysis of the structure-function relationship between TMV CP and HR in N' tobacco. Accordingly "the key step in HR induction resides with the host cell's recognition of the pathogen elicitor". Having developed an array of TMV mutants with specific amino acid subsitutions in their CP it was possible to define structural variations that impinged upon HR elicitation. *It became apparent that substitutions that elicited HR were resident along interface regions between adjacent viral subunits within ordered aggregates of CP.* Subsitutions which were HR- were either conservative or were located outside the interfacing regions. *In vitro* analysis of CP aggregates revealed that HR-eliciting substitutions reduced aggregate stability. In fact, the intensity of HR was correlated with the ability of amino acid substitution to interfere with (diminish) aggregate stability. Clearly, the HR-eliciting substitutions altered the normally tight subunit-to-subunit interactions and disrupted the formation of CP aggregates. This study also revealed that certain amino acid substitutions that altered side chain conformation could disrupt overall tertiary structure which was required for host recognition and hence is integral to HR elicitation.

In summary, HR in N ' tobacco reflects a response to a "loosening" of TMV CP quartenary structure while overall tertiary structure is maintained. The loosening effect apparently exposes a receptor binding site that is normally inaccessible for host recognition. The precise structural configuration of the CP that triggers HR remains obscured. However, the molecular size of the protein elicitor, as judged from in vitro suppression of large aggregate formation, suggests that it is a monomer or low number multimer.

The reports by Culver, Dawson and colleagues clearly implicate coat protein components and their configuration in the induction of HR. *However, this is in relation to HR necrosis in N. sylvestris (N' gene).* Supportive evidence for viral coat protein as an inducer of HR was reported by Pfitzner and Pfitzner (1992). In their study the ToMV (tomato mosaic) coat protein gene was introduced via *Agrobacterium tumefaciens* into isogenic lines of *Nicotiana tabacum* nn (N' minus) and *N. tabacum* EN (N' plus). Strong necrogenesis (HR) was expressed in calli emerging from the N' containing EN genotype, but not in calli from *N. tabacum* nn. Thus, there is clear evidence from another virus species that coat protein acts as an inducer of HR and in an additional species of *Nicotiana*. This research also reported the apparent unrelatedness of pathogenesis related protein (PR) PR-1 to HR.

Potato virus X (PVX) strains are classified on the basis of their interaction with the potato HR genes Nb and Nx. Further, the HR response of Nx potato is also determined by the coat protein gene of PVX and perhaps by PVX coat protein *per se*. Kavanaugh *et al.* (1992) suggest that it is possible that an RNA sequence in PVX elicits HR directly. However, attention is called to the finding that this is not the case with the HR-inducing TMV25 strain in *N. sylvestris,* (Culver *et al.* 1989, 1991).

Elicitor for N gene expression of HR:

The situation as it applies to the N gene in *N. tabacum* cv. Xanthi nc is clearly different. No mutant of TMV was, until recently, known to alter the necrotic reaction to a systemic one (to break HR-resistance). The report by Takamatsu (1987) reveals the fact that TMV mutants lacking the entire coat protein gene are capable of inducing HR in *N. tabacum.* Using TMV cDNA constructs that are almost coat protein free, they established that coatless TMV RNA produced necrotic lesions in *N. tabacum* cv. Samsun NN as large as those of wild-type TMV. They concluded that coat protein and its coding sequence appear to play no positive role in restricting viral multiplication or cell-to-cell movement, both integral features of local lesion necrosis. The mutants described by Takamatsu reflect the existence of separate inducers for HR in N and N' tobacco and almost assuredly different receptors as well.

This has recently been validated by the report of Padgett and Beachy (1993) that the genome of Ob (Obuda), a tobamovirus, overcomes N gene mediated HR. The Ob genome contained at least four open reading frames (ORF's) which encode for a 126 kD polypeptide with a 183 kD readthrough product, a 30.6 kD movement protein (MP) and an 18 kD coat protein (CP). They established that the MP did not contain the HR-breaking information. However, chemical mutagenesis of a full Ob clone produced a mutation in the 126 kD protein that caused a complete loss of the HR-breaking capability. It would appear, therefore, that a single ORF in the TMV genome is now available for detailed analysis of the gene that is involved in N gene-mediated HR. Sequence analysis revealed a C-to-T transition governing a proline to leucine change. Insertion of this single amino acid alteration converted the Ob TMV from a systemic infection to local lesion development in Xanthi NN tobacco.

Hence, in summary, it was shown that CP is not involved and it appears that the MP gene is also not involved in breaking N gene resistance. Padgett and Beachy used chemical mutagenesis to produce the Ob mutant, Ob NL-1, that is capable of inducing HR, maps to the 126 kD ORF, and has been sequenced. Changes in electrical charge of some proteins have been offered as factors in resistance breaking by other tobamoviruses (Meshi *et al.*, 1988; Calder and Palukaitis, 1992). However, the proline-to-leucine change does not alter charge, but may nevertheless alter the structural conformation of the protein. According to Padgett and Beachy, a computer-assisted analysis of the change predicted that two separate α helices in the wild-type protein would be replaced by a single long α helix. A hypothesis proposed as a *modus operandi* was that the altered protein might exert its influence on viral replicase affecting the rate of replication or the expression of viral RNAs. Continued mutagenesis of the Ob mutant is in progress.

THE BACTERIA-INDUCED HYPERSENSITIVE REACTION

Historical Aspects and General Considerations

The hypersensitive reaction (HR) in response to plant pathogenic bacteria was first recognized by Klement and coworkers in 1964, fifty years after the discovery of fungal hypersensitivity (Stakman, 1915). An inoculation technique developed by Klement (1963) enabled easy infiltration of plant tissues with a known concentration of bacteria and was instrumental in demonstrating that plant pathogenic bacteria also induce a hypersensitive response in resistant host and non-host plants. Klement, Farkas and Lovrekovich (1964) tested 22 different pseudomonad species or pathovars including the tobacco pathogen (compatible) *Pseudomonas syringae* pv. *tabaci* and five saprophytic species in tobacco leaf tissue. Of these, the non-host pathogens injected at a concentration of $> 10^6$ cells/ml induced tissue necrosis within 12 to 24 h and subsequent tissue desiccation, symptoms that are indicative of HR. *These symptoms were precisely confined to inoculated tissues*. Inoculation of saprophytic bacteria did not result in necrosis; rather, concentrations of 10^8-10^9 cells/ml caused weak chlorosis to develop. Similarly, when *P. s.* pv. *tabaci* was introduced into its host, tobacco, HR symptoms did not develop. However, in the latter case, bacteria multiplied and colonized the intercellular space. With advancing time bacteria spread into the uninoculated tissues causing the slowly developing, progressively enlarging, necrotic lesions of bacterial disease.

In the hours following inoculation with an HR-inducing pathogen (depending on the surrounding relative humidity), tissue flaccidity increases as turgor is lost. The tissue wilts, and later collapses completely. Tissues desiccate, often to dryness, within 24 hours. The bacterial population declines, and the surviving bacteria remain confined to the site of inoculation. Only tissue inoculated with live incompatible bacteria is killed while the remainder is spared (Fig. 41), if the water transport is not interrupted by the collapsed tissue (Pike and Novacky, unpublished observations). Multiplication of bacteria *per se* is not required for HR induction, as shown with nalidixic acid which inhibits bacterial multiplication but does not prevent HR (Meadows and Stall, 1981). Although most phytopathogenic

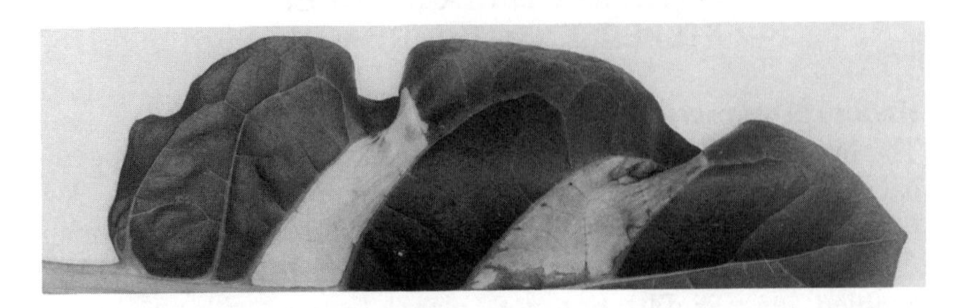

Fig. 41. Bacterial HR in tobacco leaf after inoculation with *P. s.* pv. *pisi* $>10^6$ cells/ml. Novacky, unpublished.

pseudomonads, xanthomonads and non soft-rot erwinias were found to be HR elicitors (Klement, 1982), the critical number of bacteria necessary for induction varies considerably. Some species may not elicit HR at all in incompatible interactions. Each bacterial pathovar can cause disease in only one or a few closely related hosts; thus bacterial pathogenicity is highly specific (Fahy and Persley, 1983). On the other hand, the HR is nonspecific. It may develop whenever any bacterial pathovar is in physical contact with cells of an incompatible host or non-host plant (Klement, 1982). Since the HR symptoms are elicited only by plant pathogenic bacteria and not by nonpathogenic saprophytes, HR in tobacco leaves or other tissues, *e.g.* bean pods (Klement and Lovrekovich, 1961), has become a standard test for plant pathogenicity of bacterial isolates.

The differences between the incompatible and compatible bacteria-plant reactions are striking and several researchers have found the bacteria-induced HR an attractive model system to study this facet of plant disease resistance, the rejection of a pathogen by a resistant plant (Klement and Goodman, 1967; Klement, 1982). Nevertheless, some bacteriologists (*e.g.* Ercolani, 1973; Rudolph, 1975) questioned the validity of the bacterially-induced HR as an indication of resistance. They claimed that the number of bacteria necessary to elicit the confluent HR necrosis, $> 2 \times 10^6$ cells/ml, was too large, far higher than the number in natural inocula. "Isn't this only a laboratory artifact that results from an unnaturally high concentration of inocula?" was a frequent comment at professional meetings. Only studies demonstrating that individual host cells react hypersensitively in tissues inoculated with low numbers of incompatible bacteria

(Turner and Novacky, 1974; Essenberg *et al.*, 1979; Holliday *et al.*, 1981) dispelled doubts about the validity of bacterial HR as a resistance phenomenon. On the other hand, even compatible interactions (*e.g. P. s.* pv. *tabaci* in tobacco) at high concentrations of bacteria may exhibit HR-like necrosis (Klement, 1982). In these instances the inoculum concentration must be increased to levels that are *ca.* 100x higher than a number sufficient to induce HR in incompatible interactions. Similarities between HR symptoms induced in an incompatible interaction and HR-like symptoms caused by a higher concentration of bacteria in a compatible interaction suggested that the two phenomena are closely related. Indeed, Klement (1982) contended that the HR is in essence an accelerated development of the compatible situation and the rapid cell death and/or related biochemical changes prevent bacteria from spreading in the tissue. The relationship between the two phenomena will become clearer in subsequent discussions of hypersensitivity and pathogenicity genes.

Relationships Between Bacteria and Plants

Bacteria on the plant surfaces

The phyllosphere of higher plants is exposed to a diverse group of epiphytic microorganisms that include saprophytic and plant pathogenic bacteria. Some of them attach to the plant surface, colonize it and survive there in a close, but non-pathological association with the plant (Leben, 1965; Hirano and Upper, 1990). Thus, plant surfaces comprise a distinct microhabitat for resident epiphytic flora (Leben, 1981). Bacteria may utilize organic and inorganic nutrients, that diffuse out of the plant surfaces (Tukey, 1970). Consequently bacteria may multiply and reach relatively high numbers on these surfaces apparently escaping recognition by the plant. Bacteria presumably cannot communicate with the epidermal cells beneath the waxy cuticle; therefore, epiphytic bacteria do not affect the plant harboring them. Bacterial cells and plant cells can truly interact only after bacteria have migrated into the intercellular space with the help of flagella (Panopoulos and Schroth, 1974) or entered the vascular system of the host (Goodman *et al.*, 1976).

The leaf intercellular space

Invading bacteria cannot make direct contact with the membrane of the living plant cell. The cell wall is the closest possible contact. All "signals" between bacteria and the host cells must pass through the cell wall. However, it is not the wall of only one individual cell that faces the bacterial microcolony. Cell walls constitute a continuum, the apoplast, that conducts water and solutes outside plasma membranes (Lauchli, 1976). The apoplast, rather than the walls of individual cells in the vicinity of bacteria is, therefore, the source of all nutrient diffusion.

Tab. 6. Growth of *Pseudomonas syringae* pv. *phaseolicola* supported by nutrients in apoplast

Parameter	Values	Units
1*.Growth yield	2.9 (10^6)	Bacteria/μg glucose equivalent
2. Nutrients ^{A}	2 (10^{-5})	μg glucose equivalent/plant cell
3. Nutrients ^{B}	5	μg glucose equivalent/cm^2 leaf
4. Diffusable nutrients ^{C}	37	μg glucose equivalent/cm^2 leaf
5. Supportable growth from A	60	bacteria/plant cell
6. Supportable growth from B	1.5 (10^7)	bacteria/cm^2 leaf
7. Supportable growth from C	4 (10^8)	bacteria/cm^2 leaf
8. Observed populations	4200	bacteria/plant cell
9. Observed populations	10^9	bacteria/cm^2 leaf

Values are presented in three columns in order to compare supportable *vs.* observed pathogen density.

*Comments:

1. Yields of *P. s.* pv. *phaseolicola* grown on minimum glucose medium represent an efficiency of 11% in agreement with published values for *P.s.* pv. *phaseolicola*. Values would be higher for bacteria with greater than 11% efficiencies.

2. Apoplast volumes were calculated on the basis of an average cell size 30x30x80 μm (7.2x10^5 μl) and 1 μm thick cell walls. With these values the cell wall represents 16% of the cell volume in agreement with 15-18% free space obtained from diffusion experiments. Estimates of organic solute concentrations of 10 mM glucose equivalents (5-8 mM carbohydrate and 2 to 5 mM amino acids, organic acids etc.) were based on values for short term (5-10 min) experiments.

3. Bean leaves were estimated to contain 17.6 μl/cm^2 (including the air space) and 2.4x10^5 cells/cm^2 based on wet weight, dry weight and bulk density experiments. Assuming that the apoplast is 16% of the volume and contains 10 mM glucose equivalents, the leaf would contain 2.8 μl apoplast/cm^2 and 5μg glucose equivalent /cm^2.

4. Using mild leaching conditions for bean leaves (a leaf still attached to the plant immersed in water) 7.5 mg of carbohydrate was recovered during a 24 h period. This represents 4.8 % of the leaf dry weight and 137 μg carbohydrate/ cm^2 leaf.

5. Calculations are based on the growth yield and available nutrients.

(Table reproduced from Hancock and Huisman (1981) with permission. For more information regarding the sources of various calculations the reader is asked to see the original review.)

The leaf-spotting bacterial pathogens begin to multiply and colonize the intercellular space of their respective hosts. The intercellular fluid can support the growth of saprophytic bacteria as well as pathogenic bacteria (Klement, 1965) regardless of compatibility or incompatibility with a given host. However, bacteria may use these favorable nutritional conditions only when they are in a compatible situation. A crucial condition for this is an absence of HR. A single compound may be the critical factor determining bacterial growth only in a rare situation; *e.g.* homoserine, a predominant amino acid in pea plants is toxic to many plant pathogenic pseudomonads, however, not to the pea-pathogenic *P. s.* pv. *pisi* (Hildebrand, 1973). Nutrients available for the developing bacterial microcolony in the intercellular space of the susceptible plant may support an increase in bacterial multiplication of several orders of magnitude in the course of a few days (Tab. 6).

Saprophytic and pathogenic bacteria and plants

Saprophytic bacteria alone do not have the capacity to initiate either pathogenesis or the hypersensitive reaction (HR). These associations are unique to plant pathogenic bacteria. Although saprophytic bacteria occasionally do enter the intercellular space and may survive there for an extended period of time (Hayward, 1974), their populations usually remain stationary, though there is one report of > 10-fold transient increase in the population of *P. fluorescens* (Morgham *et al.*, 1988). They apparently lack the genetic information to cause either pathogenesis or HR as will be discussed later. They multiply rarely and only under unusual conditions in which the natural resistance of a host plant is disrupted or pathogenesis is potentiated as in the following examples.

1) When a plant is coinoculated with its compatible pathogen and a saprophyte, the latter may also multiply. When Young (1974a) coinoculated *P. s .* pv. *lachrymans* and *P. fluorescens* into bean leaves, the multiplication rates of both bacteria were very similar on the first day. Similarly, populations of *P. fluorescens* increased to the level of the cotton-pathogenic *Xanthomonas campestris* pv. *malvacearum* following coinoculation of cotton leaf tissue and subsequent maintenance in continuous dark (Morgham *et al.*, 1988).

2) When plant resistance to *P. s.* pv. *lachrymans* was suppressed, *e.g.* with cycloheximide, *P. fluorescens* colonized and enzymatically degraded potato tuber (Zucker and Hankin, 1970).
3) When tobacco plants were kept in total darkness for 3 days at 100% relative humidity *P. fluorescens* multiplied. An HR-like necrosis resulted after returning to low humidity (Lovrekovich and Lovrekovich, 1970).

However, avirulent and saprophytic bacteria can alter the behaviour of pathogenic species, suppressing the infection process. For example, rapidly growing apple shoots preinoculated with an avirulent isolate of *Erwinia amylovora* or a yellow *Erwinia*-like isolate were protected against infection by a virulent strain of *E. amylovora* (Goodman, 1967). Similarly, inoculation of pear blossoms with *E. herbicola* 24 h before inoculation with a virulent strain of *E. amylovora* protected the trees from infection (Riggle and Klos, 1972). Riggle and Klos suggested that the consumption of organic nitrogen compounds by *E. herbicola* changed the intercellular pH to a level inhibitory to *E. amylovora*. Considering what is known today, it is likely that nutritional considerations affected gene induction (*e.g. hrp* genes) in *E. amylovora* (see page 179). Infection of rice leaves by *X. oryzae* was suppressed by preinoculation with *E. herbicola* (Hsieh and Buddenhagen, 1974). However, efforts to establish protective saprophytic bacterial populations have not been successful. For example, Crosse (1965) attempted to change the proportion of *P. s.* pv. *mors-prunorum* to *P. fluorescens* on cherry trees. *P. s.* pv. *mors-prunorum* outnumbered saprophytic bacteria 3-4 : 1. Spraying bacterial suspensions initially increased populations of the saprophyte. These, however, rapidly declined and the natural ratio was reestablished in a few days.

Role of bacterial extracellular polysaccharides.

Intercellular space, the site of bacterial colonization, is filled with air under natural conditions. However, there is probably a water film at the plant cell surface. Bacteria have developed the capacity to adsorb water and maintain a high water content in the intercellular space by the production of extracellular polysaccharides (EPS) in order to survive and colonize the tissue (Rudolph *et al.*, 1989). Pseudomonads and xanthomonads typically cause watersoaked lesions at infection sites (Fahy and Persley, 1983). Bacterial multiplication has been correlated with the high water content in the intercellular space that is visible as a watersoaked lesion, *e.g.* on bean leaves infected with *P. s.* pv. *phaseolicola* (Rudolph, 1984) or on leaves of different hosts infected with pathovars of *X. campestris* (El-Banoby and Rudolph, 1979b). In the absence of water soaking pathogenesis does not develop. This phenomenon has been interpreted as a resistance reaction by Rudolph and coworkers (1989).

Rudolph (1978) demonstrated that production of EPS is the mechanism by which bacteria induce water soaking. He induced persistent water-soaked spots in bean leaves with bacteria-free exudate preparations collected from infected bean leaves. Similar results were obtained with EPS preparations from bacterial cultures grown *in vitro* (El-Banoby and Rudolph, 1979a). However, these findings were not readily repeated by others, *e.g.* Fett *et al.* (1986), who tried, unsuccessfully, to induce water soaking with EPS. The relationship between the EPS-induced water soaking and the host specificity of the EPS-generating bacteria is also unclear (Rudolph *et al.*, 1989). Additionally, during the first 24-48 h following inoculation of highly resistant and susceptible cotton cultivars with *X. c.* pv. *malvacearum*, more EPS-derived sugars were produced in the incompatible combination (Pierce *et al.*, 1993).

These inconsistencies are probably a result of the complex nature of the bacterial EPS. The EPS of gram-negative bacteria exists in several physical states and is known to undergo temperature-related viscosity changes in solutions that suggest temperature-induced transitions between ordered and disordered conformations (Morris *et al.*, 1977). Within the plant host, EPS interactions with plant cell wall polysaccharide may change the viscosity of EPS and form gels. This has been observed with xanthans and plant galactomannanan and also with other polysaccharides that have the β-1,4 linked backbone (Morris *et al.*, 1977).

One report of Rudolph and coworkers on EPS as a crucial factor in bacterial pathogenesis (Klement *et al.*, 1987) is worth mentioning. When plants were in total darkness both before and after inoculation, the sugar content in the apoplast was drastically reduced, the content of alginate in infected leaves decreased dramatically, and the EPS production was 20 times lower than in illuminated leaves. Watersoaking was absent and necrotic symptoms developed. The authors suggested that the symptoms became HR-like because the decreased EPS production brought bacteria into closer contact with the host cells, a condition for the HR elicitation. This idea is in line with observations of suppression of HR by, for example, a protein lipopolysaccharide complex (Mazzucchi and Pupillo, 1976). However, considering what we know today of the relationship between nutritional status and the expression of genes controlling pathogenicity and/or HR, it is more plausible to suggest that the conditions in Rudolph's experiments affected gene expression *(e.g. hrp* genes), (see page 179).

Multiplication of bacteria

The ultimate level of the bacterial population depends on the number of bacterial cells that initiate the infection, and on host/pathogen compatibility. An increase of several orders of magnitude is commonly observed in 1-5 days in compatible combinations. For example, the population of *P. s.* pv. *phaseolicola* increased from an inoculum of 10^3 bacteria/cm^2 in bean leaves to 5 x 10^7 bacteria/ cm^2 leaf in 5 days with water soaking symptoms appearing on the fourth day. Similarly, *P. s.* pv. *morsprunorum* increased from 10^3 bacteria/cm^2 in cherry leaves to 10^7 bacteria/cm^2 leaf in 5 days, also displaying water soaking on the fourth day (Ercolani and Crosse, 1966). However, when these pathogens were inoculated at the same 10^3 cells/ml concentration on non-host plants, *i.e. P. s.* pv. *phaseolicola* in cherry leaves and *P. s.* pv. *morsprunorum* to bean leaves, numbers not only did not increase to more than 10^5 bacteria/cm^2 leaf but in addition the plants remained macroscopically symptomless. Clearly there are large differences in replication rates of bacteria in compatible and non-host incompatible combinations.

In a more detailed study of multiplication of compatible and incompatible bacterial pathogens, Daub and Hagedorn (1980) followed the change in populations of *P. s.* pv. *syringae* in susceptible and resistant bean pods and leaves in parallel with that of *P. s.* pv. *coronafaciens*, a non-pathogen of bean. In pods the differences in replication and final populations was most profound at the low inoculum level of 10^3 cells/ml. In the incompatible combinations the doubling time was 9.7 h with the non-host/pathogenic bacterium and 5.7 h with the host/non-pathogenic (avirulent) bacterium. It was only 2.3 h with the host/pathogenic (virulent) bacterium, *i.e.* in the compatible combination. With higher inoculum levels the differences among the three combinations were less divergent. At 10^9 cells/ml there were almost no differences in doubling times.

In leaves the pattern was different. In both susceptible and resistant bean leaves, *P. s.* pv. *syringae* multiplied without a lag phase and with doubling times of 1.6 h and 1.8 h, respectively. The growth rate in the two combinations became divergent only after 12 h when multiplication in the resistant host drastically declined (Fig. 42). Exponential growth in the non-host/pathogenic *P. s. pv. coronafaciens* began only after an 8-12 h lag period, and

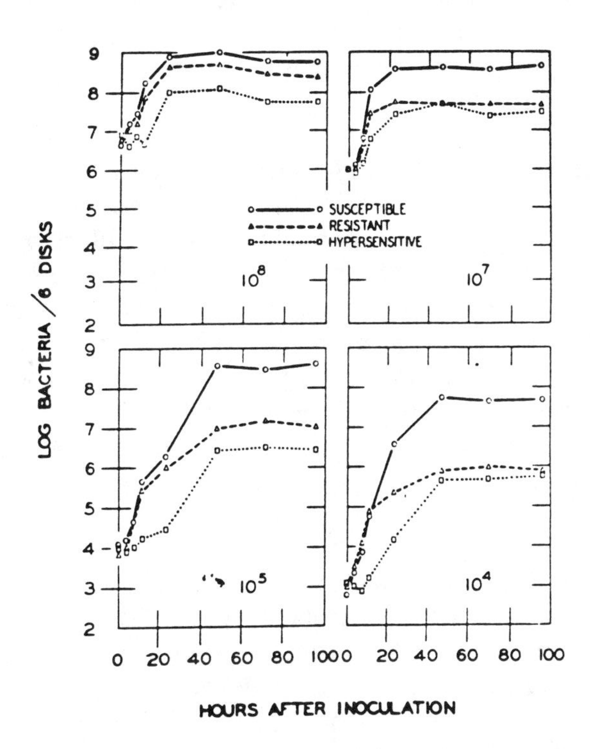

Fig. 42. Multiplication of bacteria in bean leaves in the susceptible *(P. s.* pv. *syringae* in cultivar Tenderwhite), resistant *(P. s.* pv. *syringae* in WBR 133), and hypersensitive *(P. s.* pv. *coronafaciens* in Tenderwhite) combinations at four different inoculum concentrations (10^8, 10^7, 10^5, 10^4 cells/ml). Daub and Hagedorn, 1980.

even later the doubling time was longer than that of *P. s.* pv. *syringae*. Bacteria entered the stationary phase at the same time in all three combinations: at 24 h when the inoculum concentrations were high and at 48 h when concentrations were low (Fig. 42). Macroscopic symptoms of HR developed only when the inoculum level was > 10^4 cells/cm^2 leaf. This study illustrates not only differences in the multiplication of bacterial populations in compatible and incompatible interactions but also suggests that possibly two types of HR exist.

A multiplication pattern similar to that in the resistant and susceptible bean leaves inoculated with *P. s.* pv *syringae* was observed recently when a susceptible and resistant cotton cultivar were inoculated with *X. c.* pv *malvacearum* (4.9 x 10^6 cells/ml) (Pierce *et al.*, 1993). At the beginning bacteria multiplied similarly in both cultivars. The difference between the two became obvious only after two days when the multiplication in the resistant tissue leveled off while that in the susceptible continued for at least 3 more days.

Nature of Pathogenesis by Leaf-Spotting Bacteria

We will focus our attention on *Pseudomonas syringae* and *Xanthomonas campestris* pathovars that cause leaf blight and spot diseases rather than other major types of bacterial diseases (Fahy and Persley, 1983) because the phenomenon of HR was discovered in these plant-pathogen interactions. However, *Erwinia amylovora* and *Pseudomonas solanacearum*, pathogens that are primarily invaders of vascular tissues will be discussed too, because they express compatibility and incompatibility in leaf tissues similar to leaf-spotting diseases.

Nutrient supply

Tab. 6. illustrates that pathogenic leaf-spotting bacteria may be well supplied with nutrients by the susceptible plant without "the coercive" action of toxic metabolites to ensure nutrient flow. Indeed, there is no evidence that low molecular weight toxins are involved in the initiation of leaf-spot diseases although some leaf-spotting pathogens produce perturbing metabolites such as extracellular enzymes and nonspecific low molecular weight toxic substances. Extracellular metabolic waste products and extracellular polysaccharide production, *etc.*, contribute to the disease syndrome. However, it is apparent that mutants that do not produce toxin can still multiply in the plant tissue, although without some symptoms, *e.g.* multiplication of *P. s.* pv. *tabaci* without the chlorosis caused by tabtoxin (Daniels *et al.*, 1988). Several tox$^-$ mutants retained the capacity to cause disease conclusively demonstrating that toxin activity is not always a prerequisite for disease symptoms (Rudolph *et al.*, 1989).

Since bacterial pathogens do not penetrate living cells and apparently many do not produce low molecular weight toxins such as those produced by fungal pathogens, the nature of the disease-causing mechanism of many plant pathogenic bacteria remains a question. We may hypothesize that the disease-causing principle is simply the simultaneous expression of pathogenicity gene products and suppression of host resistance genes. This combination ensures compatibility and conditions for rapidly multiplying, and therefore highly metabolically active, bacteria to colonize the plant tissue.

Pathogenicity genes are linked to genes that induce hypersensitivity commonly referred to as the *hrp* gene cluster (Lindgren *et al.*, 1986). The question concerning *hrp* gene expression in pathogenesis has been studied in the compatible association of bean infected with *P. s.* pv. *syringae* (Atkinson and Baker, 1987a). The authors used the weakly virulent strain 61 and the virulent strain Y30 of *P. s.* pv. *syringae* as well as its three Tn*5* insertion mutants with impaired ability to induce HR and K^+ efflux to determine whether the K^+ efflux also occurs during pathogenesis and how bacterial multiplication is related to it. They observed that the potassium efflux occurred during both HR development and pathogenesis. However, the increased K^+ efflux and concomitant extracellular alkalinization induced by a wild type strain during pathogenesis occurred about 2 h later than the increased K^+ efflux during HR (Fig. 43). One of the three mutants used in these experiments showed similar but less pronounced K^+ efflux.

An increase in potassium efflux preceded the onset of bacterial growth. Therefore, the authors have considered the K^+ efflux and extracellular alkalinization to be closely related to the multiplication of bacteria and to the development of foliar symptoms. Atkinson and Baker (1987b) suggested that the K^+ efflux and alkalinization play a role in providing nutrients for

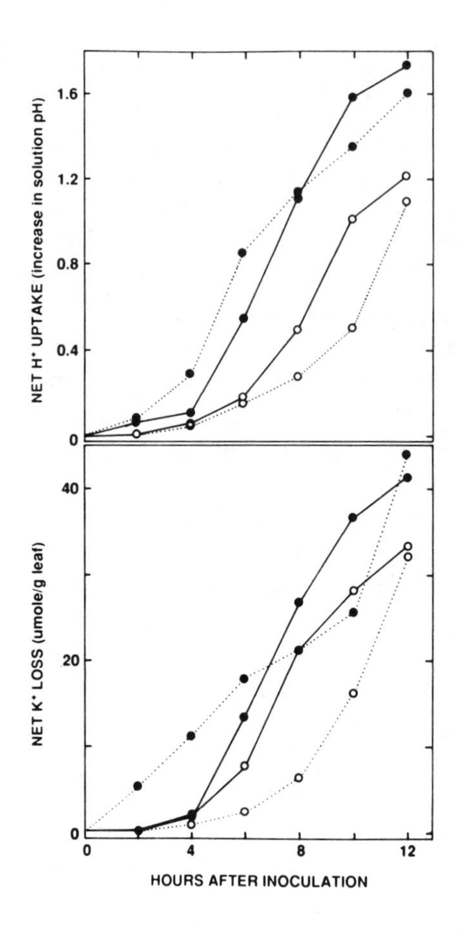

Fig. 43. Effect of *P. s.* pv. *syringae* weakly virulent strain 61 (●) and virulent strain Y30 (O) on net H^+ and K^+ transport in leaf disks from moderately resistant bean cultivar Pinto 111 (solid line) and susceptible cultivar Tendercrop (dotted line). Atkinson and Baker, 1987a.

bacterial growth. An increased extracellular pH could affect the H^+ gradient across the plasmalemma ($\Delta\mu H^+$) and in this way inhibit the uptake of sucrose via H^+/solute cotransport. This then would lead to a higher extracellular sucrose concentration. Indeed, the authors found an increased efflux and decreased influx of sucrose in bean leaf tissue 2-4 h after inoculation with *P. s.* pv. *syringae*. They further tested their hypothesis by mimicking the alkalinization of intercellular space with the infiltration of a buffer. Buffers at pH 7-8 induced alterations in sucrose transport similar to those caused by bacterial infection. These data, however, may require some reevaluation. Leaf mesophyll cells export rather than import sugars. Sucrose uptake occurs mainly at the phloem plasmalemma, the site of phloem loading (Giaquinta, 1983), not at the plasmalemma of leaf mesophyll cells. Also, it has been reported, that mesophyll cells exhibit solute (amino acids) cotransport during HR development in cotton cotyledonary tissue inoculated with *P. s.* pv. *pisi* (Novacky, 1980).

Permeability changes during bacterial pathogenesis

The assumption that invading plant pathogens induce changes in the host cell permeability is shared widely by plant pathologists (Wheeler, 1975). However, evidence for this is rather limited in bacterial pathogenesis. From the few studies on the permeability changes during bacterial pathogenesis it is worth mentioning that an increase in the size of water soaked lesions in cucumber leaves infected with *P. s.* pv. *lachrymans* was paralleled by increased permeability to solutes, detected as the ^{42}K efflux (Williams and Keen, 1967). Also, during the exponential growth of leaf-spotting bacteria (*P. s.* pv. *phaseolicola* in bean leaves) there was an increased efflux of ^{35}Sulfur loaded before inoculation (Mobley *et al.*, 1972). Similarly Burkowicz and Goodman (1969) found increased electrolyte leakage induced in apple leaf tissue by both virulent and avirulent strains of *E. amylovora*.

Membrane potential studies

How then does the developing microcolony affect the host plant and how do bacteria damage cell membranes and eventually kill the tissue? We will attempt to answer this question by discussing the electrophysiological membrane studies of a typical leaf-spot disease, the angular leaf spot of cotton, caused by *X. c.* pv. *malvacearum* (Novacky and Ullrich-Eberius, 1982). The electrical membrane potential (E_m) was measured with glass microelectrodes (Higinbotham, 1973; Standen *et al.*, 1987) in studies of HR and pathogenicity because this technique allows detailed examination of cell membrane function *in situ* soon after bacteria have established contact with plant cells (Novacky *et al.*, 1976). Measurements of the electrical membrane potential (E_m) with glass microelectrodes give a reliable picture of cell membrane transport activities. They are performed on tissue sections with only minimal disturbance. Both the energy-independent, passive (ion diffusion) component and the active membrane transport component, *i.e.* electrogenic pump, properties may be identified with this technique. When the ATP flow to the plasmalemma is interrupted under conditions that inhibit its synthesis, for example, in chloroplasts in the dark and in mitochondria under the influence of metabolic inhibitors, *e.g.* cyanide, azide or anoxic conditions, the plasmalemma H^+-ATPase is inhibited and the E_m components are separable. Such an interruption is recorded as a membrane depolarization, specifically, a loss of the energy-dependent component of membrane potential. The remaining potential, the energy-independent component is identified as the passive diffusion potential.

When cotton cotyledonary tissue (cv. Acala 44) was inoculated with *X. c.* pv. *malvacearum* race 10 (a compatible combination) no alterations were observed in E_m during the first 20 h after inoculation. However, at 24 h post-inoculation and later, E_m values were similar to the healthy control only in the light. In the dark, E_m rapidly decreased to the level of the diffusion potential (Fig. 44). Nevertheless, the energy supply (ATP), photosynthetic O_2 evolution and respiratory O_2 uptake in light and dark were not affected (Novacky and Ullrich-Eberius, 1982). The reason for this E_m anomaly in dark conditions must be a change(s) in some cellular process that can be ameliorated by photosystem II; for example, an enzyme system that would reduce the toxic amounts of ammonia previously reported to occur during bacterial pathogenesis (Lovrekovich *et al.*, 1970; Ullrich *et al.*,

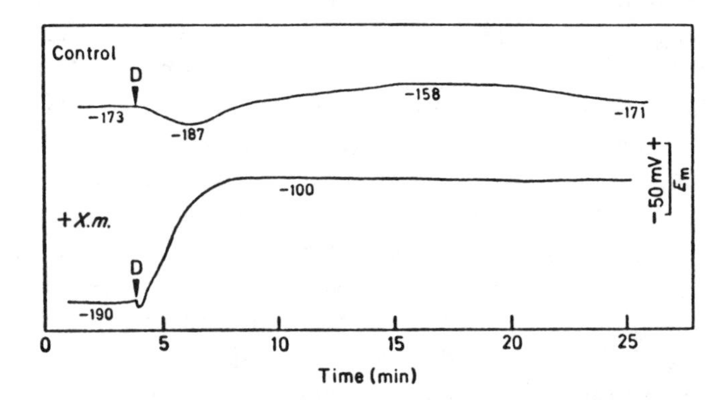

Fig. 44 . Membrane potential (E_m) in light and darkness in healthy control and in *X. c.* pv. *malvacearum* (X.m.) inoculated cotton cotyledon segments. Numbers on traces denote recorded mV. Novacky and Ullrich-Eberius, 1982.

1984). The photosystem II-dependent glutamine synthetase-glutamate synthase cycle [GS-GOGAT cycle] (Miflin and Lea, 1982) may be the process involved.

Ullrich-Eberius *et al.*, (1989) suggested that ammonia produced by bacteria during the development of cotton blight is assimilated by the photosystem II-dependent GS-GOGAT cycle in the light. However, in the dark, when photosystem II is no longer operative, NH_4^+ begins to accumulate in the host cells. This increased NH_4^+ concentration may alkalinize the cytosol and lead to inactivation of the plasmalemma and/or tonoplast H^+-ATPase(s) resulting in membrane depolarization in the dark. These collective changes may subsequently lead to alterations in primary cell metabolism, catabolism of storage products (another source of NH_4^+) and solute efflux.

Several authors (*e.g.* Goodman, 1972; O'Brien and Wood, 1973) rejected the idea of Lovrekovich *et al.* (1970) that ammonia produced by bacteria is a critical necrotrophic factor in bacterial pathogenesis and HR. They contended that the increased ammonia level during bacterial pathogenesis originates from protein degradation and therefore is a late nonspecific consequence of the tissue injury. However, bacteria produce ammonia in culture and also during the initial phase of pathogenesis (Stall *et al.*, 1972; Lovrekovich and Lovrekovich, 1970; Lovrekovich *et al.*, 1970; Ullrich *et al.*, 1984). Although this initial production may not be related to later-developing tissue necrosis, it may have a critical role in the elicitation of host cell responses, especially the increase in membrane permeability (Ullrich-Eberius *et al.*, 1989). Ammonia accumulates to a high level in the dark and less in the light. Treatment of cotton tissue with a similar concentration of ammonia depolarized E_m immediately to the level of the diffusion potential (Ullrich *et al.*, 1984). It is important to remember that at much lower concentrations of ammonia than that used in E_m experiments by Ullrich *et al.* (1984), the cytosol is alkalinized by up to 0.5 pH unit (Bertl *et al.*, 1984).

Nature of Bacteria-Induced Hypersensitive Reaction

Requirement of living bacteria for HR induction

Both the disease syndrome and the HR have been believed to be induced only by living bacteria that are capable of protein and RNA synthesis (Klement and Goodman, 1967; Sequeira, 1975; Sasser, 1978, 1982; Meadow and Stall, 1981). The prerequisite of protein synthesis for the HR development was well illustrated by the experiments of Ercolani (1970) using a histidine-requiring mutant of *P. s.* pv *syringae* on a non-host tomato plant. The HR developed only after infiltration with histidine. A strong increase in uptake of uridine-H^3 by *P. s.* pv. *pisi* immediately after inoculation also indicates that protein synthesis is elevated, since *ca.* 50% of the newly synthesized RNA is ribosomal RNA (Sigee and Al-Issa, 1982).

However, the presence of living bacteria is crucial only during the initial phase of contact between bacterium and plant cell surface, thus strongly suggesting a transfer of the HR-eliciting molecules, *e.g.* harpin (see page 179), at this time. The HR induction period became apparent in detail when inoculated tissues were treated with prokaryotic-specific antibiotics at various intervals after inoculation. The longest elapsed time between inoculation of bacteria and an antibiotic treatment that prevented development of HR was designated the induction period (Klement and Goodman, 1967; Sequeira, 1975; Sasser, 1982). Subsequent antibiotic treatment had no effect on later phases of HR development and the final outcome, necrosis. The induction period varies among different plant/bacteria combinations from 30 min for tobacco/*P. s.* pv. *pisi* (Klement and Goodman, 1967), 60-90 min for bean/*P. s.* pv. *morsprunorum* (Lyon and Wood, 1976), 120 min in tobacco/*P. solanacearum* (Sequeira, 1975) to 4 hr in bean/*P. s.* pv. *phaseolicola* (Roebuck *et al.*, 1978). A correlation between bacterial multiplication and the HR induction was established by Somlyai *et al.* (1988).

The HR induction period could be the time that is necessary for certain (*e.g. hrp*) genes to be expressed, *i.e.* the time needed to synthesize (*e.g. hrp*, see page 178) gene products. Xiao *et al.* (1992) reported enhanced expression of *hrp* genes 2 h post-inoculation. Seven of the eight *hrp* gene units could be controlled by environmental factors during an incompatible interaction. Similarly, Yucel *et al.* (1989) changed the time of HR induction by changing the nutritional conditions under which the bacteria were grown (see page 172).

The experiments also clearly showed that the inhibition of HR by antibiotics is a result of action on the bacterium and not on the plant. Sasser (1982) evaluated the inhibition of HR by 50 antibiotic compounds. Only those that inhibit bacterial protein and RNA synthesis were effective in suppressing HR development. When bacterial mutants were selected for their resistance to eleven antibiotics which effectively inhibit HR development, HR resulted in every case that the bacterial mutant and corresponding antibiotic were injected into tobacco leaves (Sasser, 1982). It is important for our disccussion to note that none of the antibiotics that inhibit the bacterial cell wall synthesis prevented HR.

Attachment and recognition of bacteria

Requirement for bacteria/plant contact: The same surface-to-surface forces that exist on the plant exterior may operate in incompatible plant-pathogen interactions in the plant interior. However, our present understanding of the attachment process comes from studies of host-pathogen interactions in which adherence to specific sites on the cell surface is an essential part of the infection process. The transfer of genetic information to a host by the crown gall bacterium, *Agrobacterium tumefaciens* (Zambryski, 1988), or the successful infection of a specific legume host root by a symbiont *Rhizobium* spp. (Pueppke, 1984) are examples. Bacterial pathogens that apparently differ from agrobacteria and rhizobia in their mode of infection, may also attach to plant cells, but in the incompatible, rather than the compatible association. Light and electron microscopic investigations strongly indicate that bacteria attach to cell surfaces in the intercellular space and that this attachment occurs more often and/or in a different way in incompatible than compatible combinations (Goodman *et al.*, 1976; Sequeira *et al.*, 1977; Érsek *et al.*, 1985). However, the evidence for attachment in HR is not unequivocal. Other authors found little correlation between HR induction and bacterial attachment (Daub and Hagedorn, 1980; Fett and Jones, 1982), or that attachment *per se* often did not lead to an HR collapse (Sequeira *et al.*, 1977).

That close contact between incompatible bacteria and plant cell is a necessary step in HR induction was demonstrated by Stall and Cook (1979). First they found that the HR did not develop if the leaves of a resistant cv. of pepper (*Capsicum annuum)* were kept water-soaked after inoculation with *X. c.* pv.

vesicatoria (Stall and Cook, 1979). Continual water-soaking did not alter the reaction in leaves of susceptible pepper cultivars as bacteria multiplied and colonized the tissue. In later studies they inoculated leaves with bacteria suspended in 0.5% water-agar. Upon gelation, bacteria were immobilized in intercellular space thus precluding contact with the resistant host cell walls and HR symptoms did not develop. This evidence may not prove that contact is required, because these tissues were not examined for HR at a microscopic or ultrastructural level. Under similar conditions, water-soaked tobacco leaf tissue developed ultrastructurally observed membrane damage during HR, although tissue desiccation and collapse was prevented by the 100% relative humidity (Goodman, 1972).

Attachment or localization of bacteria: The attachment and/or localization phenomenon has been thoroughly investigated in plant/xanthomonad complexes. Al-Mousawi and coworkers (1983) examined two incompatible situations in cotton (*Gossypium hirsutum*) leaves: *X. c.* pv. *malvacearum* in a resistant cotton (cv. Im. 216) and *X. c.* pv. *campestris* in cv. Acala 44 as well as a compatible interaction *X. c.* pv. *malvacearum* in cv. Acala 44. They included a gram-positive bacterium *Micrococcus lysodeikticus,* heat-killed, UV-killed and rifampicin-killed cells of *X. c.* pv. *malvacearum*, corn starch granules and polysterene latex particles in their controls. Both bacteria that induced HR (*X. c.* pv. *malvacearum* in cv. Im. 216 and *X. c.* pv. *campestris* in Acala 44) and all killed *X. c. malvacearum* cells were localized. Acala 44 enveloped the gram-positive bacteria and starch granules. On the other hand, *X. c.* pv. *malvacearum* was free of envelopment in the compatible cultivar Acala 44. Latex particles were never enveloped in either cultivar. The authors concluded that envelopment is not a specific recognition phenomenon, but rather the result of the hydrophilicity of particles and bacteria. Latex particles may have escaped envelopment because they are hydrophobic. Only compatible bacteria avoided envelopment, perhaps by the rapid synthesis of EPS. However, the most recent study of cotton bacterial blight concluded that EPS could not play a role in preventing resistance responses to *X. c.* pv. *malvacearum* because equal quantities of EPS are produced in susceptible and resistant cotton lines (Pierce *et al.*, 1993).

These studies did not examine the moment of attachment *vis à vis* the time of the HR induction, but rather the localization or immobilization phenomenon of incompatible bacteria *per se* as a component of the HR process. Therefore, it has not yet been established that the attachment of incompatible bacteria is requisite for HR induction (Sequeira, 1985). Jones and Fett (1985) investigated four incompatible strains of *Xanthomonas campestris* on soybean leaves and concluded that "immobilization may be a generalized response of the mesophyll cell to the sustained presence of live bacteria at the surface" but does not play a role in the development of the HR.

Hildebrand *et al.* (1980) related the ultrastructural localization of compatible or incompatible *P. s.* pv. *syringae* strains, the incompatible *P. marginalis*, *P. s.* pv. *pisi*, *P. s.* pv. *tomato*, and the saprophytic *P. fluorescens* in bean (*Phaseolus vulgaris*) leaves to the evaporation of intercellular water. Cell wall fibrils, according to them, originate from cell wall components separated during the initial infiltration of water and to the subsequent evaporation of water. Therefore, they concluded that the envelopment of bacteria is a passive phenomenon rather than a recognition event. Their assessment, however, does not account for several features revealed by Goodman *et al.* (1976), Sequeira *et al.* (1977), Politis and Goodman (1978), Cason *et al.* (1978), and Jones and Fett (1985): wall and wall-membrane modifications and entrapment of bacteria on cell wall surfaces other than junctions between cells.

The most recent study of fibrillar material presents an opposite view. Brown *et al.* (1993) investigated the origin of a fibrillar matrix around cells of *X. c.* pv. *vesicatoria* during HR and pathogenesis in pepper tissues immunocytochemically with monoclonal antibodies specific for the xanthan side chain of EPS. They demonstrated unequivocally that the fibrillar material surrounding *X. c.* pv. *vesicatoria* cells is bacterial EPS, rather than plant wall components.

Sequeira (1985) discussed the possible role of plant hydroxyproline-rich glycoproteins in the attachment of incompatible *P. solanacearum* in *Nicotiana tabacum*, but it has not been demonstrated immunologically. The specific attachment of *P. s.* pv. *glycinea* races to solitary leaf cells of resistant soybean cultivars has been attributed to plant lectin and bacterial carbohydrate binding since attachment could be inhibited with particular sugars in a cultivar-dependent manner (Érsek *et al.*, 1985).

Role of bacterial surface molecules: The question of which molecules on the surfaces of bacteria and host cells are involved in the recognition process has been addressed but at this time has not been answered. Sequeira (1985) performed a series of experiments with an array of *P. solanacearum* strains that are wilt-inducing and have a diverse host range. The pathogen produces large quantities of extracellular polysaccharide (EPS) that is associated with pathogenicity (Kelman, 1954). Strains that lose the capacity to produce EPS, *e.g.* by mutation in culture, lose virulence. When EPS production is lost or considerably reduced, the lipopolysaccharide formation also is altered. Though the lipid A region and the R core region of LPS are identical in both strains, the entire O-polysaccharide region is missing in the avirulent strain (Whatley *et al.*, 1980; Baker *et al.*, 1984; Sequeira, 1985).

Sequeira *et al.* compared K60, a virulent strain that produces EPS (EPS^+) and does not induce HR (HR^-), with B1, an avirulent mutant that produces no EPS (EPS^-) and induces HR (HR^+). These two strains behaved very differently in the intercellular space of tobacco. The K60 strain freely multiplied in and colonized the intercellular space, whereas the avirulent mutant B1 attached to the mesophyll cell wall surface and its multiplication was restricted (Sequeira *et al.*, 1977). Several other properties of the avirulent B1 strain were found to be different from the wild type, such as decreased cellulase activity (Kelman and Cowling, 1965), increased motility (Kelman and Hruska, 1973) and increased ability to deaminate and decarboxylate tryptophane (Phelps and Sequeira, 1967). Perhaps, the most interesting difference between virulent and avirulent strains was the distribution of pili; the surface of avirulent cells was covered with pili whereas very few pili were on the surface of the virulent stain. However, Stemmer and Sequeira (1984) found no structural difference in pili isolated from the avirulent strain and the pili of other Gram-negative bacteria.

Using hybridoma cell lines, monoclonal antibodies have been raised against the cell surface of pathogenically characterized *P. s.* pv. *glycinea* races including strains that contain the *avrA* gene (Wingate *et al.*, 1990). Analysis of the bacterial cell surface, however, did not reveal a correlation between either monoclonal antibody-recognized surface components or the expression of the *avrA* gene and the behaviour of bacterial strains in soybean cultivars.

Expression of bacteria-induced HR

Ultrastructural characteristics: The most prominent ultrastructural characteristics of HR induced by *P. s.* pv. *pisi* in tobacco during the first 2-3 h after inoculation are cell wall

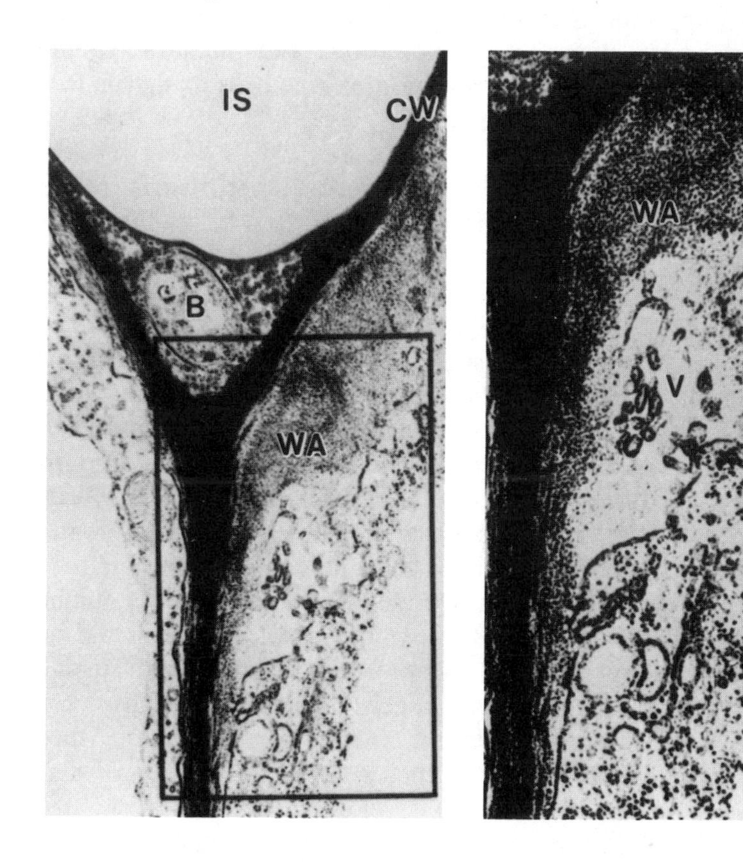

Fig. 45. Bacterial cells (B) of *P. s.* pv. *pisi* embedded in a mixture of amorphous and microfibrillar material at a cell junction (CW, cell wall). By 4 hr after inoculation wall appositions (WA) are well formed structures associated with numerous vesicles (V) (inset). (x17,000, inset: x22,500). Politis and Goodman, 1978.

thickening, separation of the plasmalemma from the wall surface, vesiculation of the plasmalemma in positions opposite to the bacteria and aggregates of microfibrils and membrane-bound vesicles in the space between the plasmalemma and the inner cell wall (Goodman *et al.*, 1976; Politis and Goodman, 1978). These aggregates, termed wall appositions (Fig. 45), have also been observed in leaves of other plants undergoing the bacterial HR (Jones and Fett, 1985; Morgham *et al.*, 1988). From 4-6 h after inoculation, the wall apposition becomes a more organized dome-shaped structure as the HR development ensues. Large numbers of membrane-bound vesicles were associated with the cell wall appositions and the plasmalemma in the same location. The formation of cell wall appositions and membrane-bound vesicles indicates an active response of host cells to signals from "invading" incompatible bacteria.

Macroscopic *vs.* microscopic HR: Resistant tissues remain symptomless if the concentration of bacteria is below the critical threshold for induction of confluent necrosis ($<5x10^6$ cells/cm^2). However, such tissues develop microscopic lesions (Turner and Novacky, 1974; Essenberg *et al.*, 1979). In fact, in some instances a single bacterium may be sufficient to elicit the HR reaction in a single plant cell as observed in tobacco leaf tissues inoculated with *P. s.* pv. *pisi* (Turner and Novacky, 1974). When a suspension of $5x10^3$ cells/cm^2 of *P. s. pv. pisi* was introduced into the intercellular space of tobacco leaf, the first dead mesophyll cells were detected as soon as 2-3 h after inoculation. At 6-7 h after inoculation the number of dead cells increased to a maximum of $4.5x10^3$ cells/cm^2 (Fig. 46) which was proportional to the number of bacterial cells inoculated. This number of dead cells did not increase with time (Fig. 47). A linear relationship with a slope of 1.0 was found between bacterial cells introduced and plant cells killed 12 h after inoculation (Fig. 48). Based on these observations it was calculated that a minimum of 25% of plant cells must be killed in order to develop confluent necrosis.

The HR at the microscopic level was demonstrated later in bean leaf tissues inoculated with *P. s.* pv. *morsprunorum* (Lyon and Wood, 1976), in cotton inoculated with *X. c.* pv. *malvacearum* (Essenberg *et al.*, 1979), and in soybean inoculated with *P. s.* pv. *glycinea* (Holliday *et al.*, 1981). In some combinations the critical

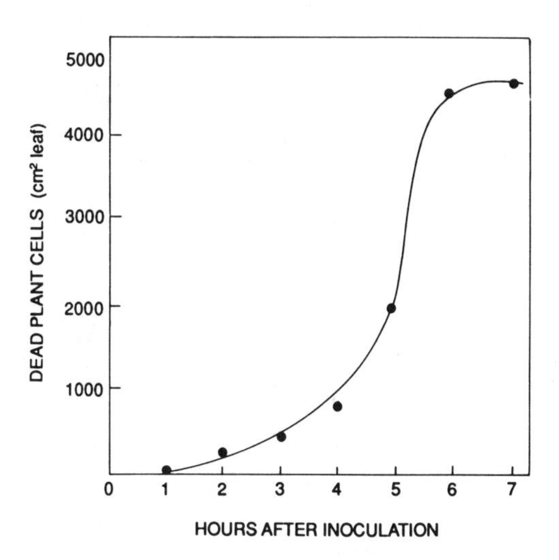

Fig. 46. Dead mesophyll cells, detected by their selective staining with a 1.0% aqueous solution of Evans blue, in tobacco leaves at intervals after the introduction of 5.0×10^3 cells/cm^2 leaf of the incompatible pathogen *P. s.* pv. *pisi*. Turner and Novacky, 1974.

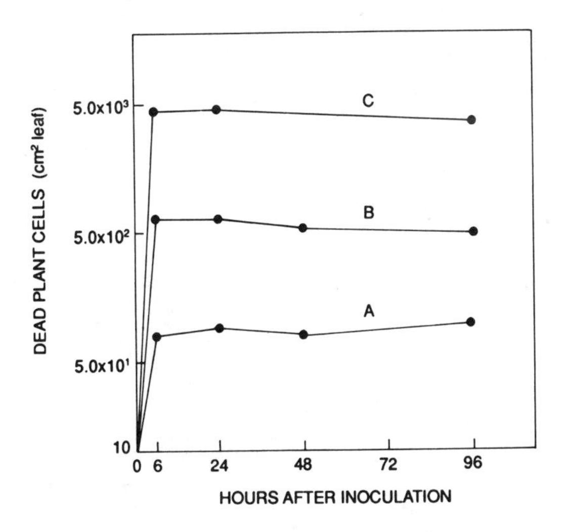

Fig. 47. Dead Mesophyll cells detected by selective staining with a 1.0% aqueous solution of Evans blue, in tobacco leaves at intervals after the introduction of 5.0×10^1 (A), 5.0×10^2 (B) and 5.0×10^3 (C) cells/cm^2 leaf of the incompatible pathogen, *P. s.* pv. *pisi*. Turner and Novacky, 1974.

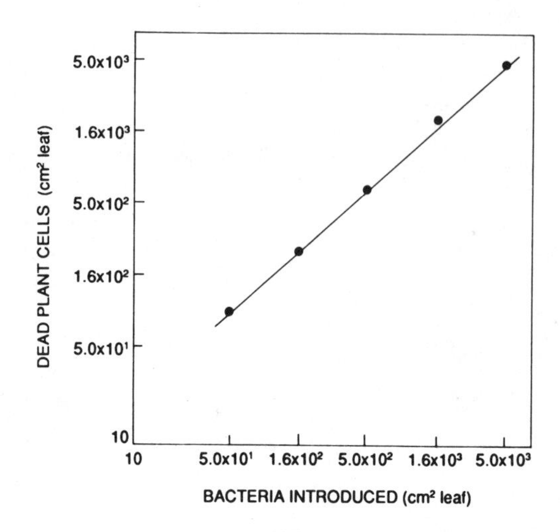

Fig. 48. Relation between the number of cells of *P. s.* pv. *pisi* introduced into leaves of the non-host tobacco and the number of dead mesophyll cells detected in those leaves 12 h after inoculation by selective staining with a 1.0% aqueous solution of Evans blue. Turner and Novacky, 1974.

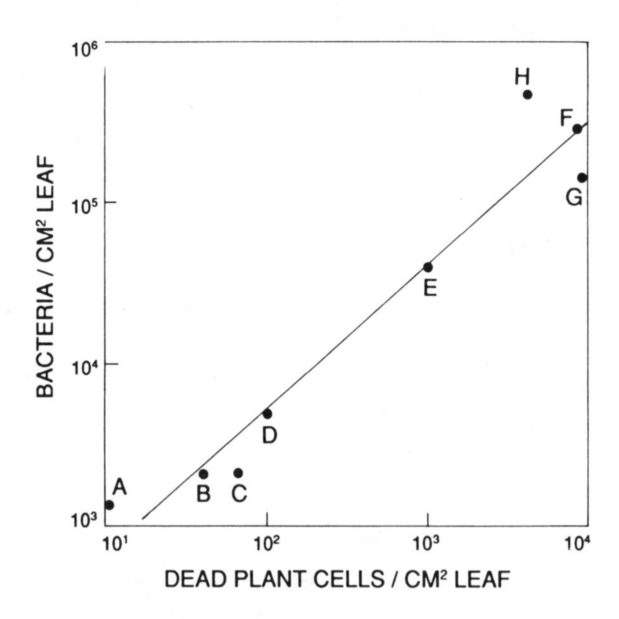

Fig. 49. Relation between bacterial multiplication and plant cell death in tobacco leaves 48 h after the introduction of an inoculum dose containing 10^5 cells/ml of *P. fluorescens* (A), *P. tomato* (B) *P. phaseolicola* (C), *E. amylovora* (D), *P. pisi* (E), *P. syringae* (F), *P. tabaci* (G) and *P. solanacearum* 60-1 (H). Turner and Novacky, unpublished.

number of bacteria required for the death of one plant cell is higher (Fig. 49) (Turner and Novacky, unpublished). It may be as high as 100 bacterial cells/plant cell as in the case of tobacco/*Pseudomonas cichorii* (Stall and Cook, 1973).

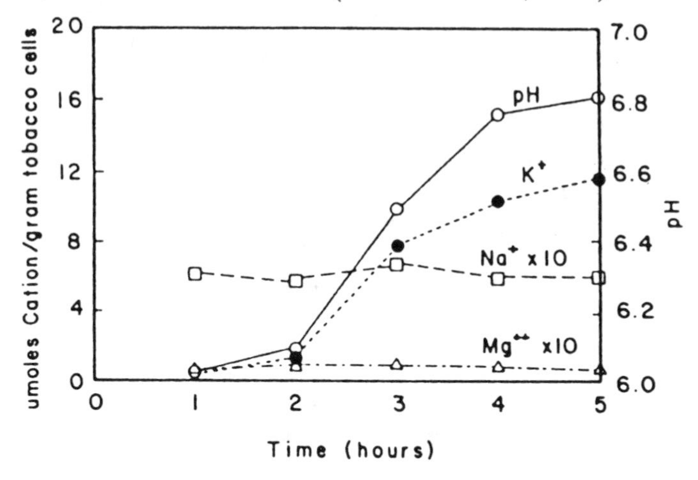

Fig. 50. Cumulative cation loss from, and increase in the assay medium pH of tobacco callus cells inoculated with *P. s.* pv. *pisi*. Data represent the concentrations of cations or the pH of the medium at intervals after inoculation of tobacco cells with bacteria. Na^+ and Mg^{2+} are shown at 10 times actual concentrations. The data for Ca^{2+} is superimposable with that for Mg^{2+} and is not shown. Atkinson *et al.*, 1985b.

Alterations of cell membrane

Electrolyte efflux: An early manifestation of membrane alterations in hypersensitively responding plant tissues is an increased leakage of electrolytes 2-3 h after inoculation with HR-inducing bacteria (Cook and Stall, 1968; Goodman, 1968; Lyon and Wood, 1976). Efflux of electrolytes, which indicates a loss of the structural integrity of cell membranes, has been related to membrane alterations found in electron microscopic studies (Goodman and Plurad, 1971; Goodman, 1972; Goodman *et al.*, 1976). Both types of studies support the contention that the plasmalemma is critically changed during HR.

Bacteria-induced electrolyte leakage was studied also in tobacco suspension cultured cells (Atkinson *et al.*, 1985b). The predominant ion in plant cells, potassium, is lost more rapidly than other ions during HR (Fig. 50) and the pH of the external medium correspondingly alkalinizes. The stoichiometric relationship found between K^+ efflux and alkalinization (1:1) prompted Atkinson and coworkers (1985b) to propose that a "K^+/H^+ exchange" mechanism operates across the plasmalemma during the HR and consequently the cell interior acidifies (Fig. 51). Such an acidification was demonstrated by Pike and coworkers (1992) by using a fluorescent pH probe (seminaphthorhodafluor) and a confocal laser scanning microscope, to follow the change in intracellular pH of cotton suspension-cultured cells treated with the non-host pathogen, *P. s.* pv. *tabaci*. Within 1 h after treatment an intracellular acidification was observed.

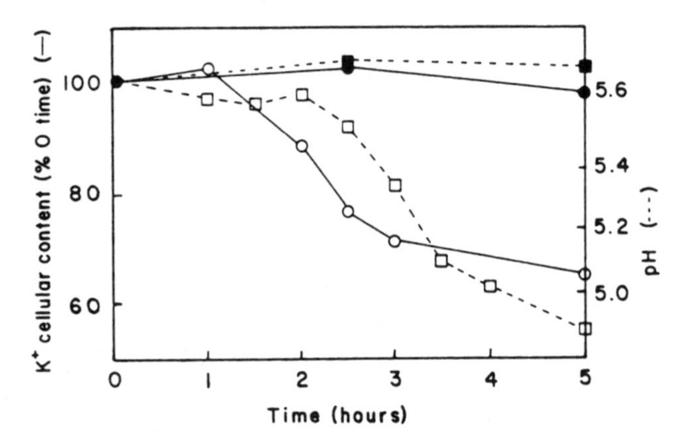

Fig. 51. Net loss of K^+ from tobacco callus cells and acidification of tobacco callus cell sap during the first 5 h of the HR following inoculation with *P. s.* pv. *pisi.* (O, □), inoculated tobacco cells: (●, ■), control tobacco cells. Atkinson and Baker, 1985b.

The H^+ transport in plant cells is under strict control of the plasmalemma H^+-ATPase, the master enzyme of the energy-dependent membrane transport (Spanswick, 1981; Serrano, 1989), and the cell's biophysical pH stat (Smith and Raven, 1979). The relationship between the activity of the H^+-ATPase and the HR-inducing K^+ efflux and external alkalinization was studied by Atkinson and Baker (1989). They treated tobacco cells with H^+-ATPase inhibitors or a protonophore 2.5 h after inoculation and found that 90% or more of both K^+ efflux and extracellular alkalinization were reduced (Fig. 52, 53). Respiratory inhibitors and inhibitors of glycolysis had a similar effect. The apparent dependence of K^+ efflux and accompanying external alkalinization on the H^+-ATPase activity suggested that the membrane transport machinery must be in an "active" operating state to be attacked by

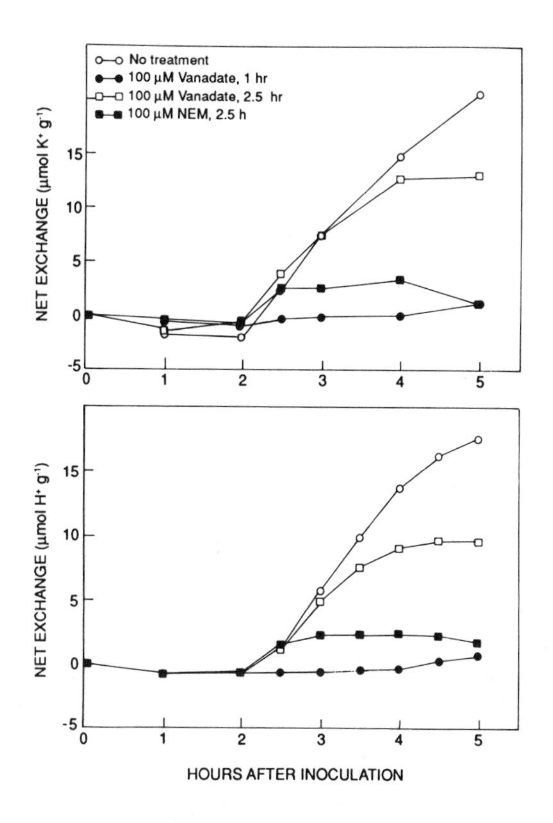

Fig. 52. Effects of the plasma membrane ATPase inhibitors vanadate and N-ethylmaleimide (NEM) on the K^+/H^+ response in hypersensitive tobacco cells. Cell suspensions were inoculated at 0 h with wild-type (hypersensitive) or HR-negative (control) strains of *P. s.* pv *syringae*. Inhibitors were added to cell suspensions at 1 or 2.5 h after inoculation. Data represent differences in net K^+ and H^+ transport between hypersensitive and control tobacco cells. Positive values indicate K^+ loss from hypersensitive cells and alkalinization of the external medium relative to controls. Atkinson and Baker, 1989.

the HR-eliciting molecules. Therefore, these experiments could be interpreted as suggesting that K^+ efflux during the early phase of HR may proceed through the K^+ channels rather than through generalized, unspecified lesions in the damaged membrane as suggested by Pavlovkin *et al.* (1986).

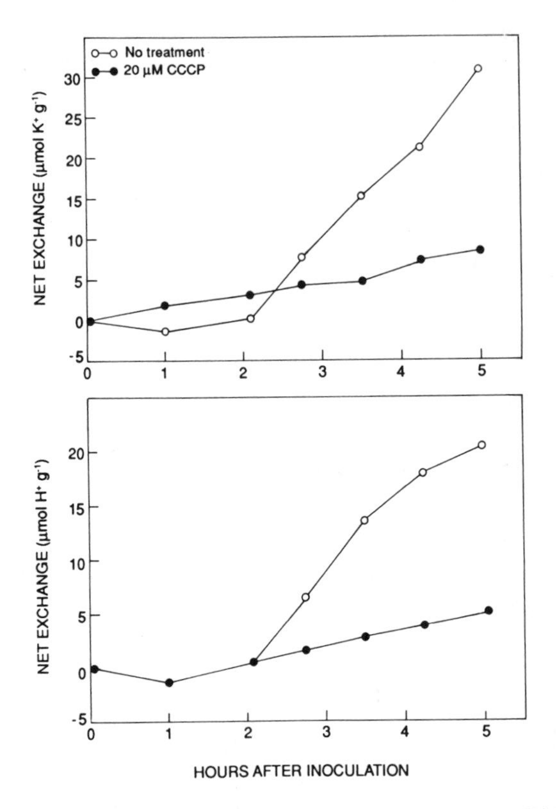

Fig. 53. Effects of the protonophore CCCP on the K^+/H^+ response in hypersensitive tobacco cells. CCCP was added to cell suspensions at 2 h after inoculation. Other details are as in the legend of the preceding figure. Atkinson and Baker, 1989.

Strong ion difference: The relationship of the extracellular alkalinization to the massive HR-related K^+ efflux is not completely clear. It may be evaluated in terms of the "strong ion difference" (SID) of a solution which determines the pH as described by Stewart (Stewart, 1981; Guern and Ullrich-Eberius, 1988; Ullrich and Novacky, 1992). The SID is the arithmetic sum of the strongly dissociated cations such as K^+ and anions as Cl^- that determines the amount of a weakly dissociated cation such as H^+ which will be present in solution. Of course, H^+ may enter cells that have lost K^+. When $K^+:H^+ = 1:1$ the external alkalinization could be sufficient evidence for an H^+ influx in exchange for K^+. Alternatively, a corresponding anion may accompany the K^+ efflux to balance the charges. The question then is: which anion is the balancing anion?

As previously discussed, the evolution of ammonia during the bacterial pathogenesis or HR and the concept of ammonia as a toxic metabolite of bacteria has been questioned (Goodman, 1972; O'Brien and Wood, 1973). Nevertheless, significant amounts of ammonia accompanying alkalinization have been detected both within and outside tissues during the HR development (Lovrekovich *et al.*, 1970; Stall *et al.*, 1972; Ullrich *et al.*, 1984; Ullrich *et al.*, 1994). Therefore ammonia may play a role in the alkalinization of the external medium in addition to the "K^+/H^+ response". Further stoichiometric measurements are needed: K^+: H^+: NH_4^+: HCO_3^- to determine the mechanism of K^+/H^+ response during the early phase of HR.

Role of calcium: Despite the crucial role of Ca^{2+} in membrane transport processes and signal transduction the role of Ca^{2+} in HR development was not closely examined until recently. Eukaryotic cells maintain a low intracellular Ca^{2+} concentration with a steep gradient across the plasmalemma. The Ca^{2+} gradient, *i.e.* 1000 μM Ca^{2+} outside the plasma membrane to 0.01 μM inside, is achieved by Ca^{2+}-ATPases that pump Ca^{2+} out to the cell exterior or into intracellular compartments and organelles (Marme, 1989). Atkinson *et al.*, (1990) reported an apparent increased Ca^{2+} uptake from a baseline level of 0.02-0.06 μmol/g/h in the control to 0.5-1.0 μmol/g/h by tobacco cells responding hypersensitively to *P. s.* pv. *syringae* 2-3 h after inoculation. This startling 10-to 50-fold increase suggests that Ca^{2+} influx increases during the first few hours of HR and that the influx is a part of the HR development. Inhibition of the K^+ efflux/alkalinization and HR cell death with calcium channel blockers, indicates that increased Ca^{2+} influx is, in fact, required for the development of bacterially induced HR. Plant plasmalemma and tonoplast Ca^{2+} fluxes through the calcium channels or via the Ca^{2+}-ATPase (Hille, 1984; Briskin, 1990) all may be affected.

Atkinson *et al.* (1990) also hypothesized that calcium plays a role in the HR induction through multiple signalling pathways similar to those in animal systems (Rasmussen and Barrett, 1984; Marme, 1989). A 20-30% decrease in phosphatidylinositol which began approximately the same time as the increase of K^+ efflux and which can be inhibited by La^{3+} (Atkinson *et al.*, 1993), suggests that plant systems may utilize Ca^{2+}-dependent phospholipases in signalling pathways similar to those in animal systems. That La^{3+} inhibits both phytoalexin induction (Stab and Ebel, 1987) and the HR (Atkinson *et al.*, 1990) suggests that the calcium ion may play a central role in plant defense reactions. Phosphatidylinositol breakdown, calcium influx (Fig. 54) and K^+ efflux were arrested by a phospholipase inhibitor, bromophenol-acylbromide (BPAB) and neomycin (NM), an inhibitor of phosphoinositide kinases (Atkinson *et al.*, 1993).

A report by Schroeder and Hagiwara (1989) that cytosolic calcium regulates voltage-dependent K^+ channel activity is important for our discussion. These authors found that elevation of cytosolic Ca^{2+} to micromolar concentrations leads to activation of a voltage-dependent anion channel and inhibition of potassium uptake channels. Such an effect of calcium on K^+ channels may be related to the membrane depolarization during HR discussed on page 147.

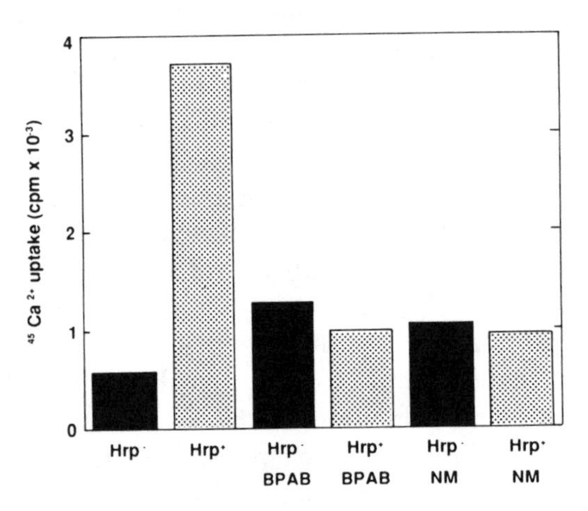

Fig. 54. Effect of inhibitors on calcium influx during the K^+/H^+ response of tobacco cell suspensions to *P. s.* pv. *syringae*. BPAB (20 μM) and neomycin (NM, 100 μM) were added to tobacco cell suspensions 1.25 hr after inoculation. ^{45}Calcium uptake was measured between 1.5 and 2.5 hr after inoculation. Modified from Atkinson *et al.*, 1993.

Electrical membrane potential (E_m): Electrophysiology, described in the section on bacteria-induced pathogenesis (see page 129), was also used to study the HR (Pavlovkin *et al.*, 1986). Membrane transport properties in cotton cotyledonary tissues were monitored after inoculation with *P. s.* pv. *tabaci* which induces HR in this plant. Membranes depolarized (Δ 35 mV) 2 hr after inoculation and became less negative as HR developed. The time needed to evaporate excess water from the cotyledonary tissue following bacteria infiltration (*ca.* 1 h) and to wash (age) tissue sections (an additional 1 h) postpones the first electrode impalement by *ca.* 2 h. Hence, it is apparent that changes may occur even much earlier than 2 h after inoculation. Indeed, an immediate transient hyperpolarization was measured following the addition of bacteria to cotton suspension cultured cells (Popham *et al.*, 1992). However, this hyperpolarization may or may not be related to membrane depolarization measured later during HR.

In cotton cotyledonary tissue, both E_m components exhibited a steady decline (Fig. 55); however, there was a difference between the two. Although the energy-dependent portion (the electrogenic pump) declined 40-60% during the first 2 h after inoculation, the activity remained constant between 2-10 hr and was detectable until the last E_m measurement (10 h after inoculation). The energy-independent potential (the diffusion potential) declined steadily to the lowest value (Fig. 55, ■), *ca.*

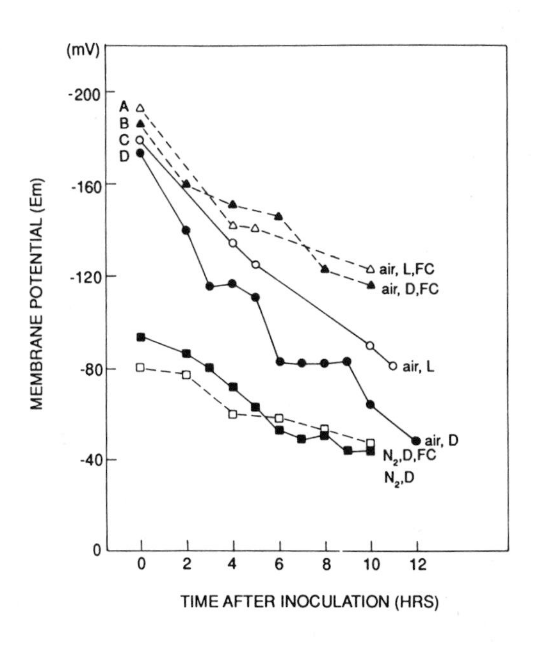

Fig. 55. Time course of the membrane potential of cotton cotyledons after inoculation with *P. s.* pv. *tabaci*, in the light (open symbols) or the dark (closed symbols), under the following conditions A and B, air + fusicoccin (△ and ▲, respectively); C and D, air (○ and ●, respectively); anoxia (N_2) (■); anoxia (N_2) +fusicoccin in the dark (□). Modified from Pavlovkin *et al.*, 1986.

45% of the control. It is important to note that the diffusion potential of a living cell is the sum of several charges. A portion of these charges is fixed, *e.g.* those in the cell wall. For example, in the study of compartmentation by Rona *et al.* (1980), the cell wall potential portion of the diffusion potential was approximately 35mV. Hence, it is detectable even in dead cells. Therefore, the 45% of the diffusion potential, 40 mV, found in the last measurable E_m during bacterial HR (Pavlovkin *et al.*, 1986) should be considered to be the potential of a dying (or dead) cell, rather than the diffusion potential of a living cell.

In parallel with the E_m measurements, the loss of electrolytes was also measured (Fig. 56), the potassium concentration [K^+] within tissues was analysed (Tab. 7) and the Nernst potential for K^+ (E_{K+}) was calculated (Fig. 57). These E_{K+} values were usually very close to the diffusion potential measured by Spanswick (1981) and correlated closely with the measured diffusion potential. A comparison of the decreased [K^+] in tissues with the sum of the effluxing electrolytes indicated that the majority of the lost ions was K^+ (Tab. 7).

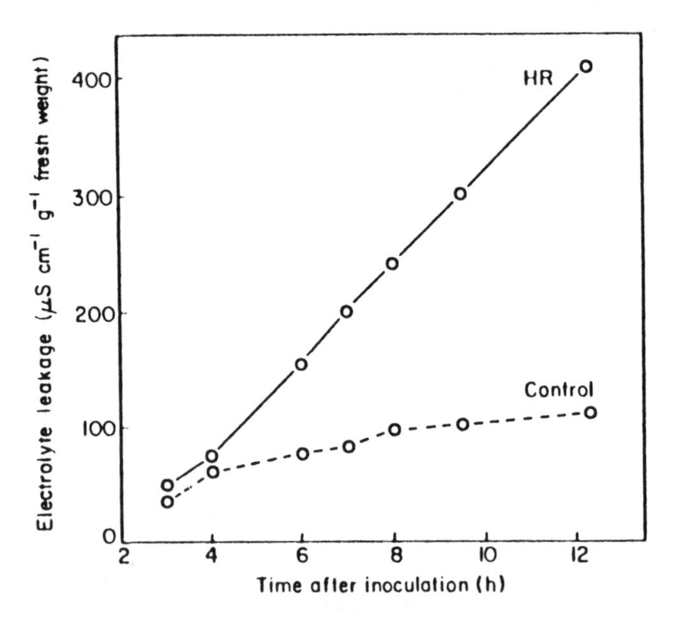

Fig. 56. Time course of electrolyte leakage of cotton cotyledon disks following inoculation with *P. s.* pv. *tabaci* (HR, hypersensitive response). Pavlovkin *et al.*, 1986.

Tab. 7. Comparison of decrease in tissue K^+ content (μmol K^+/g fresh weight) during electrolyte leakage into distilled water or 1 mM $CaSO_4$ as incubation medium[a].

Cotton	Medium	Time after inoculation (h) 1	12	24
Control	H_2O	45	42 (59%)[b]	43 (0·14%)
	$CaSO_4$		42	42(0·08%)
HR[c]	H_2O	48	28·8(82·0%)	4·8 (34·1%)
	$CaSO_4$		29·7	4·2 29·2%

[a]Measurements were made at 1, 12 and 24h after inoculation of cotton cotyledons with 10^8 cells/ml of *P. s.* pv. *tabaci*. (hypersensitive response), dark.
[b]Percentages in parentheses are %K^+ of the sum of electrolytes leaked out into the supernatant.
[c]HR, Hypersensitive response. Pavlovkin *et al.*, 1986

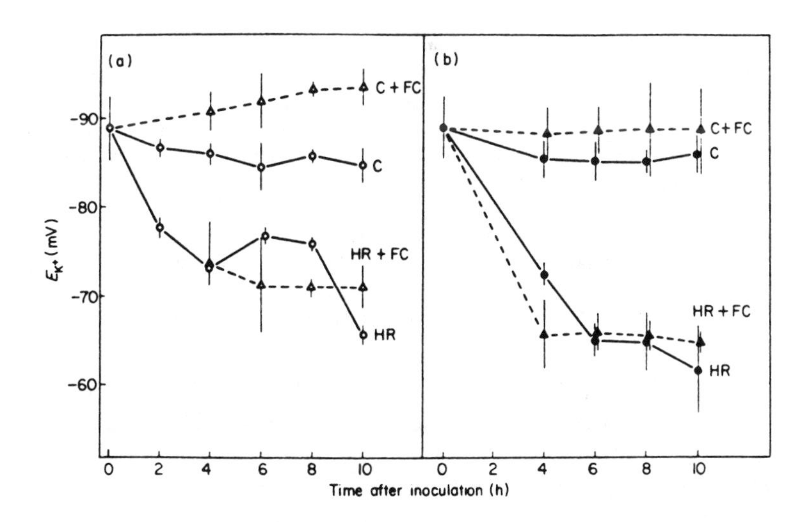

Fig. 57. The Nernst potential of K^+ (E_{K+}) of cotton cotyledons following inoculation with *P. s.* pv. *tabaci* (HR, hypersensitive response: C, control), in the absence or presence of fusicoccin (FC) in light (a) and dark (b). Pavlovkin *et al.*, 1986.

Similar alterations in the active and passive components of E_m were observed both in pepper cv. 10R, resistant to *X. c.* pv. *vesicatoria* (Fischer and Novacky, 1965), and in the incompatible combination of cucumber with *P. s.* pv. *pisi* (Keppler and Novacky, 1986). However, in resistant cotton cultivars inoculated with the pathogenic bacterium *X. c.* pv. *malvacearum,* a decrease in the diffusion potential was not detected at 10 h after inoculation (Novacky *et al.*, 1976; Pike and Novacky, unpublished). This difference most likely reflects the slower pace of HR in the resistant host as compared to the non-host.

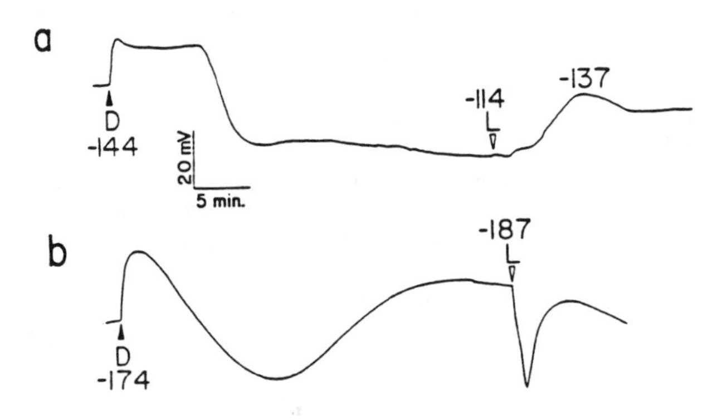

Fig. 58. Segments cut from Im216 cotton cotyledons injected with *X. c.* pv. *malvacearum* 10^8 cells/ml (a) or water (b) were submerged in an aerated bathing medium and measured more than 24 h after inoculation. As compared to the water-injected control: E_m hyperpolarization in the bacteria-inoculated cotyledon was extended after turning off the light (D), and the subsequent depolarization did not return to the resting potential; after turning on the light (L), depolarization was absent. Pike and Novacky, unpublished.

The probability of measuring cells that are in direct contact with the bacteria is rather low, because only a small portion of cells is affected by bacteria directly. This is evident from the observations of cell death during HR (Turner and Novacky, 1974; Essenberg *et al.*, 1979; Holliday *et al.*, 1981). Therefore, most cells measured by microelectrodes are not directly affected by bacteria. Hence, depolarized membranes most likely reflect the transfer of electrical signals from affected to non-affected cells via plasmodesmata. Membrane depolarization throughout the tissue could be a signal to healthy cells to synthesize products such as phytoalexins which, though they accumulate in dead cells in bacteria-infected tissue (Pierce and Essenberg, 1987), may be synthesized in neighboring living cells (Essenberg, personal communication).

When tissue collapse was prevented by incubating it in the experimental solution, reexamination of E_m the next day revealed that, while many cells were dead, a portion of cells was alive with a diffusion potential repolarized to approximately that of healthy cells (Pike *et al.*, 1991; Novacky *et al.*, 1993). These "recovered" cells, however, exhibited alterations in E_m light/dark responses (Fig. 58). These alterations may very well be a result of accumulated phytoalexins or other products that are toxic to both plant cells and bacteria.

Lipid peroxidation

The dramatic decline of the electrical membrane potential during the HR, especially in the energy-independent potential, accompanied by a loss of K^+, led to the postulation that membranes are structurally changed. Lipid peroxidation could account for structural membrane alterations. Therefore, possible alterations in membrane lipids during HR development were investigated (Keppler and Novacky, 1986; 1987).

Lipid peroxidation is a consequence of conditions in which membrane deterioration results from the oxidative perturbation of cell membranes. It has been found during leaf senescence (Fig. 59) (Dhindsa *et al.*, 1981), wounding (Theologis and Laties, 1981; Thompson *et al.*, 1987), anoxia (Hunter *et al.*, 1983), herbicide injury (Chia *et al.*, 1981), pollution (Elstner, 1982) and drought (Dhindsa and Matowe, 1981). Lipid peroxidation of cell membranes involves formation of activated oxygen species and organic free radicals from their reaction with unsaturated fatty acids (Freeman and Crapo, 1982).

This process can be initiated by the oxygen free radicals, superoxide, ($O_2^{\cdot-}$), and/or hydroxyl ($OH^{\cdot}$), generated from O_2 via reduction processes. It can also be initiated by lipid radicals and superoxide generated enzymatically by lipoxygenase (Pryor, 1976). Both superoxide and hydroxyl radicals can peroxidize unsaturated fatty acids. Although a basal level of activated oxygen species exists even under optimal conditions in cells of living organisms, their level is strictly controlled by the enzymes superoxide

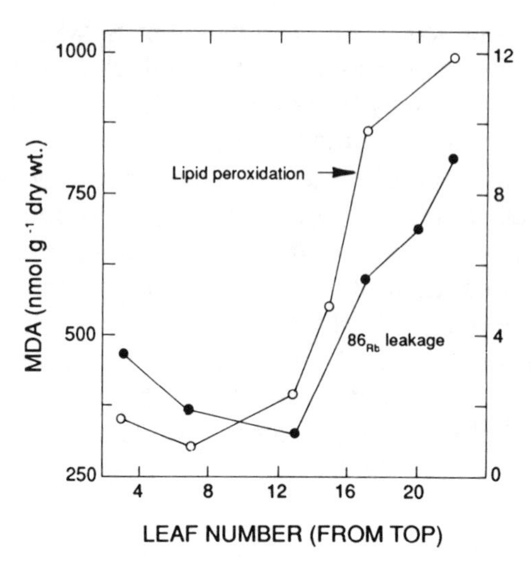

Fig. 59. Changes in lipid peroxidation (MDA content, O) and in the rate of ^{86}Rb leakage (●) during growth and *in situ* senescence of tobacco leaves. Leaves in position 13 and greater were fully expanded. Dhindsa *et al.*, 1981.

dismutase (SOD) and catalase (Freeman and Crapo, 1982). Several *in situ* antioxidants such as glutathione, ascorbic acid, α tocopherol, dipyridamole and polyamines have been reported to diminish membrane lipid peroxidation (Thompson *et al.*, 1987). Decreased activities of SOD, catalase, lipoxygenase and/or the intracellular level of antioxidants may cause an increased level of activated oxygen species and thus influence the rate of lipid peroxidation. Interference with the normal operations of a wide spectrum of biological processes themselves (Freeman and Crapo, 1982) may lead to a sudden increase in the level of active oxygen, seen as the "oxidative burst" (Apostol *et al.*, 1989b), and result in lipid peroxidation.

An increase in lipid peroxidation, measured by monitoring the conversion of lipids to malondialdehyde, is the most pronounced membrane damage caused by oxygen free radicals observed during the HR. Lipid peroxidation was found in cucumber cotyledons between 1.5 and 3 hr by Keppler and Novacky (1986) or between 3 and 4 hr in tobacco by Ádám and coworkers (1989) after inoculation with the incompatible pathogens, *Pseudomonas s.* pv. *pisi* or *syringae*, respectively. Maximum lipid peroxidation was detected at 4.5 hr (*P. s.* pv. *pisi*) (Fig. 60) or 3.5 h (*P. s.* pv. *syringae*) after inoculation (Fig. 61).

Increases in lipid peroxidation were compared with increased electrolyte leakage. A measurable increase in electrolyte loss during the HR in cucumber cotyledons was found between 3 and 4 h (Fig. 62) and in tobacco between 4.5 and 5 h (Fig. 61) after inoculation, *i.e.* later than the increase in lipid peroxidation. This strongly suggests that the lipid peroxidation precedes the increased K^+ efflux.

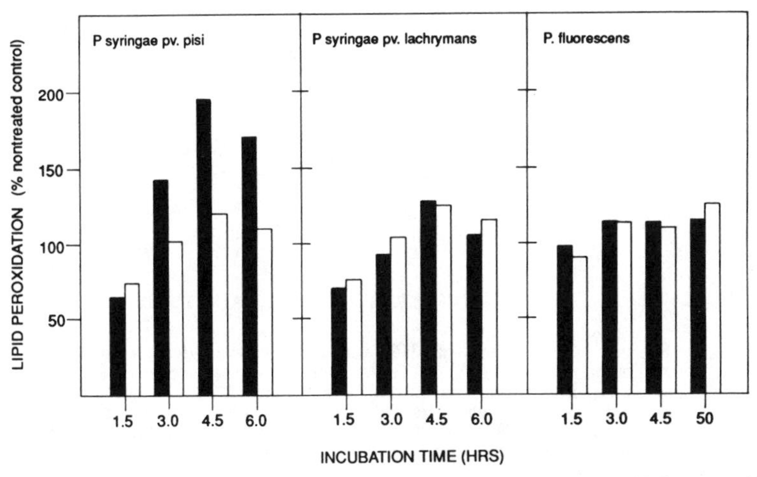

Fig. 60. Lipid peroxidation of cucumber cotyledons infiltrated with *P. s.* pv. *pisi* (incompatible combination), *P. s.* pv. *lachrymans* (compatible combination), or *P. fluorescens* (saprophyte), live or heat-killed (10^8 cells/ml), expressed as percent of nontreated controls (assayed in parallel). Solid bars = live bacterial treatments, open bars = heat-killed bacterial treatments. Keppler and Novacky,1986.

A marginal peroxidation increase was found also in the cucumber cotyledons injected with heat-killed cells of the HR-inducing bacterium, *P. s.* pv. *pisi* as well as in cotyledons inoculated with the live and heat-killed compatible pathogen, *P. s.* pv. *lachrymans* (Keppler and Novacky, 1986). The living or heat-killed saprophytic bacterium, *P. fluorescens,* did not induce lipid peroxidation at all. This strongly indicates the presence of heat stable eliciting molecules within plant pathogenic bacteria which are released by autoclaving. Eliciting molecules released by autoclaving might be identical with a novel bacterial protein, harpin (Wei *et al.*, 1992a, see page 180). If so, such harpin-like molecules must be produced below the HR-eliciting concentration, because heat-killed bacteria injected into the plant tissue do not cause confluent necrosis.

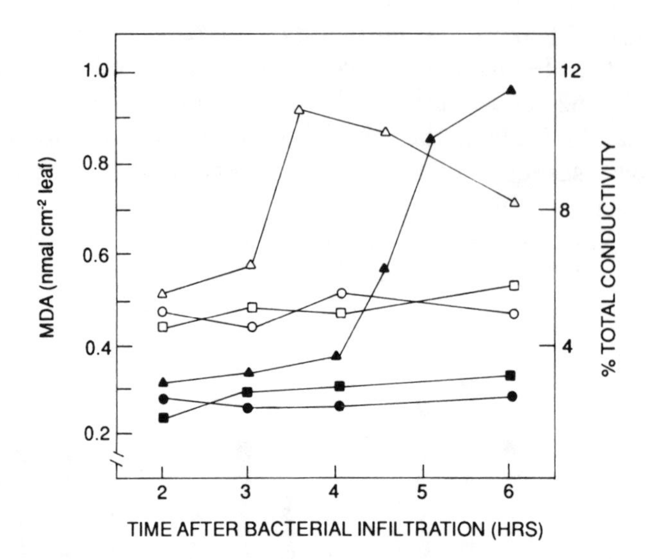

Fig. 61. The effect of HR on lipid peroxidation measured by the malondialdehyde (MDA) test (open symbols) and on electrolyte leakage (closed symbols). Leaves were infiltrated with water (O, ●), *P. s.* pv. *syringae* (10^8 cells/ml) (Δ, ▲), or with the Tn5 transposon mutant of *P. s.* pv. *phaseolicola* (10^8 cells/ml) (□, ■) which had lost the capacity to induce an HR. Electrolyte leakage was expressed as a percentage of total ion content of leaf disks. Ádám *et al.*, 1989.

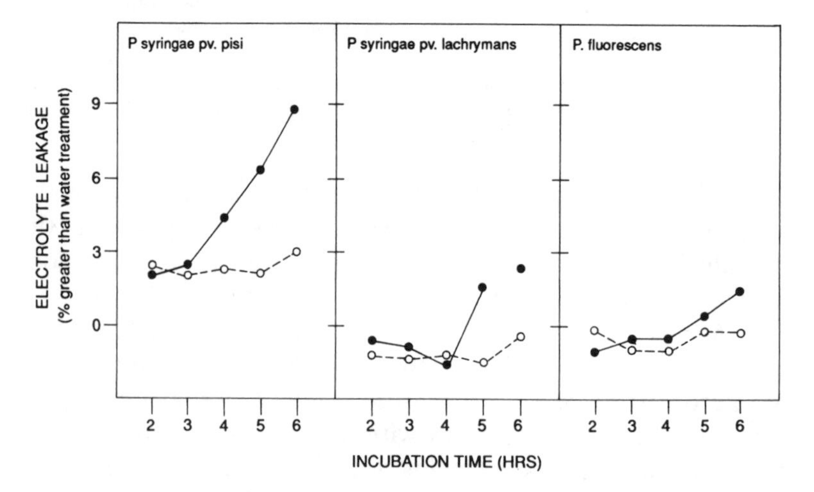

Fig. 62. Electrolyte leakage of cucumber cotyledons infiltrated with water or *P. s.* pv. *pisi* (incompatible combination), *P. s.* pv. *lachrymans* (compatible combination), or *P. fluorescens* (saprophyte), live or heat-killed (10^8 cells/ml). Electrolyte leakage of bacteria-infiltrated cotyledons expressed as mean ± SE of percent total conductance (μmhos) after autoclaving minus mean of percent total conductance after autoclaving of water-infiltrated control. ● = live bacterial treatments, O = heat-killed bacterial treatments. Keppler and Novacky, 1986.

The role of superoxide ($O_2^{\cdot-}$) in lipid peroxidation has been suggested by several authors. Ádám and coworkers (1989) monitored the $O_2^{\cdot-}$ level during the HR development by nitroblue tetrazolium (NBT)-reducing activity and found that NBT-reducing activity increased (Fig. 63) about 3 h after inoculation of tobacco with *P. s.* pv *phaseolicola*. The NBT-reducing activity was not increased when an (HR^-) Tn*5* transposon mutant was used as a control. The $O_2^{\cdot-}$ generation was followed by an increase in lipid peroxidation. Two scavengers of $O_2^{\cdot-}$, superoxide dismutase (SOD) and tiron (4,5-dihydroxy-1,3-benzene-disulfonic acid) inhibited

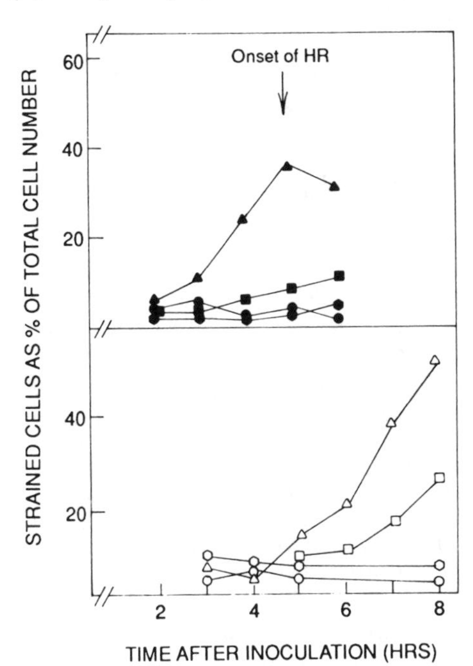

Fig. 63. Leaf disks of tobacco cv. Burley 21, were stained with (a) 300 μM nitroblue tetrazolium (NBT) solution or (b) 0.25% (w/v) Evans blue solution. The leaf disks had been infiltrated with water (●, ○), *P. s.* pv. *syringae* (10^8 cells/ml) (▲, △), *P. s.* pv. *syringae* (10^8 cells/ml) and SOD (100 μg/ml) (■, □), or with the Tn5 transposon mutant of *P. s.* pv. *syringae* (10^8 cells/ml) (♦, ◇) which had lost the capacity to induce HR. Ádám *et al.*, 1989.

but did not eliminate lipid peroxidation. Finally, treatments with SOD (Keppler and Novacky, 1987), SOD, tiron, ascorbic acid, glutathione, α-tocopherol-acetate and albumin (Ádám *et al.*, 1989) (Fig. 64) decreased the rate of electrolyte leakage suggesting that $O_2^{\cdot-}$ initiates lipid peroxidation during the HR. It is interesting to note that albumin is known to eliminate free radicals in mammalian blood (Halliwell and Gutteridge, 1986). Attempts to detect other species of active oxygen, such as singlet oxygen (Salzwedel *et al.*, 1989) and hydroxyl radical (Popham and Novacky, 1991) were negative. However, their role in HR can not be ruled out, because these free radicals are so reactive that they may escape detection.

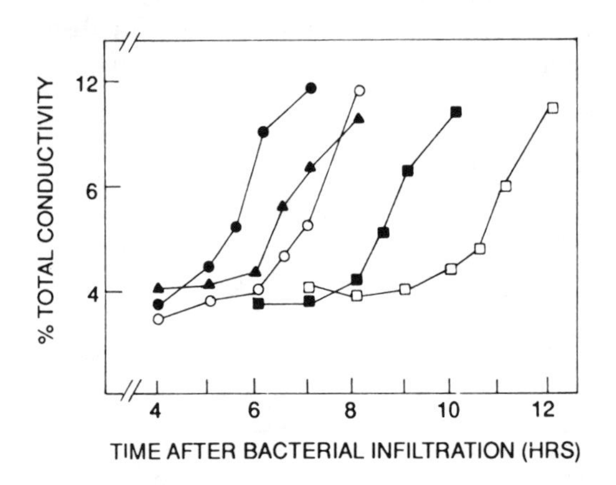

Fig. 64. The effect of GSH (100 mM), α-tocopherol-acetate (0.75 mM) and BSA (0.5-1.0% w/v) treatments on the development of *P. s.* pv. *syringae* induced HR. The time course of electrolyte leakage was measured after bacterial infiltration (5 x 10^6 cells/ml). The leaves had been treated with 0.05 M phosphate buffer (pH 7.8) (●); GSH (▲); α-tocopherol-acetatc (O); 0.5% BSA (■) or 1.0% BSA (□). In the case of BSA treatments the bacterial concentration was 10^7 cells/ml. The results are expressed as percentage total ion content. Ádám *et al.*, 1989.

Suspension cultured cells exhibit increases in free radical production and lipid peroxidation similar to those found in tissues. In tobacco cell suspension cultures increased lipid peroxidation (Fig. 65) and superoxide production induced by *P. s.* pv. *syringae* was detected 2 h after addition of bacteria (Keppler and Baker, 1989). Increases in a lipid peroxidation product, malondialdehyde, and the evolution of ethane resulting from the lipid peroxidation process, paralleled superoxide production (measured as NBT-reduction). A Tn*5* insertion mutant of *P. s.* pv. *syringae* that does not induce HR did not cause lipid peroxidation, $O_2^{\cdot-}$ production or K^+ efflux.

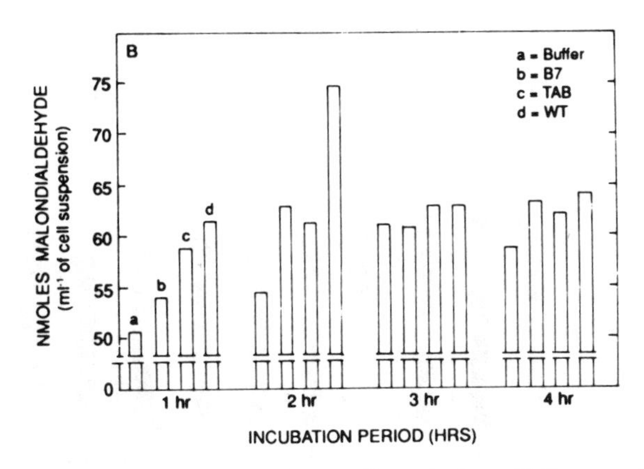

Fig. 65. Lipid peroxidation in tobacco cell suspensions. Cell suspensions were exposed to bacteria (final concentration of 10^8 colony-forming units/ml) in MES buffer (1.0 mM, pH 6.0) or buffer alone (a). Bacteria used were: B7 (Tn5 insertion mutant of wild type; does not induce hypersensitive reaction), (b); *P. s.* pv. *tabaci* (TAB) (compatible pathogen of tobacco) (c); or wild type (WT) (*P. s.* pv. *syringae* strain 61, induces HR) (d). Keppler and Baker, 1989.

Role of lipoxygenase in lipid peroxidation: In addition to the peroxidation effects of non-enzymatically activated oxygen species, lipid peroxidation may also be accomplished enzymatically by lipoxygenase (LOX) (Thompson *et al.*, 1987). Keppler and Novacky (1987) demonstrated that lipoxygenase is involved in the HR process when they detected a significant increase in lipoxygenase activity 4.5 hours after inoculation of cucumber cotyledons with HR-inducing *P. s.* pv. *pisi* (Fig. 66). Neither *P. s.* pv. *lachrymans* (a compatible pathogen), nor a saprophyte, *P. fluorescens*, altered the lipoxygenase activity. However, the increased activity of lipoxygenase was detected later than lipid peroxidation. The authors concluded that lipoxygenase activation during HR development is a plant response to a sudden increase in the level of free fatty acids after non-enzymatic peroxidation of phospholipids and is not the primary cause of HR.

A different situation was found in French bean (*Phaseolus vulgaris*) infected with host-pathogenic and host-nonpathogenic races of *P. s.* pv. *phaseolicola* (Croft *et al.* (1990). In the incompatible combination the authors detected an increased activity of two lipoxygenase isozymes 3-4 h after inoculation, that coincided with the HR-induction period in the French bean (Roebuck *et al.*, 1978). This increased lipoxygenase activity

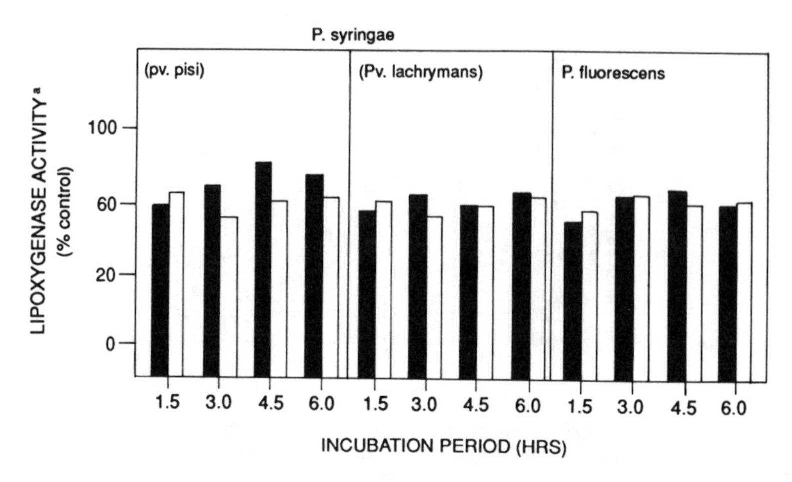

Fig. 66. Lipoxygenase activity in infiltrated cucumber cotyledons. Activities are expressed as percentages of nontreated controls (assayed in parallel). ■, live bacterial treatments; □, heat-killed bacterial treatments. Keppler and Novacky, 1987.

preceded the evolution of ethane, the final breakdown product of fatty acid hydroperoxides, that was used as an indicator of lipid peroxidation (Konze and Elstner, 1978). Croft *et al.* (1990) concluded that the lipoxygenase activity was a result of *de novo* protein synthesis because in parallel experiments cycloheximide inhibited both the lipoxygenase activity and HR development (Fig. 67). Based on these results the authors considered the lipid peroxidation found in tissue undergoing HR to be initiated by lipoxygenase. Once initiated by lipoxygenase, the lipid peroxidation would propagate autooxidatively in membranes *in situ*. According to this scheme the active oxygen species are generated as a consequence of lipoxygenase activity.

However, fatty acids must be released from the membrane phospholipids before lipoxygenase can oxidize them. It has been suggested that the enzyme lipolytic acyl hydrolase is the releasing agent (Thompson *et al.*, 1987). Indeed, Croft *et al.*, (1990) found that lipolytic acyl hydrolase increased in parallel, but not clearly prior, to an increase in lipoxygenase activity during the development of HR in French beans inoculated with *P. s.* pv. *phaseolicola*. A close examination of their data shows that while the activity of LOX-l and LOX-3 increased 4 and 6 h after inoculation respectively, the activity of lipolytic acyl hydrolase was

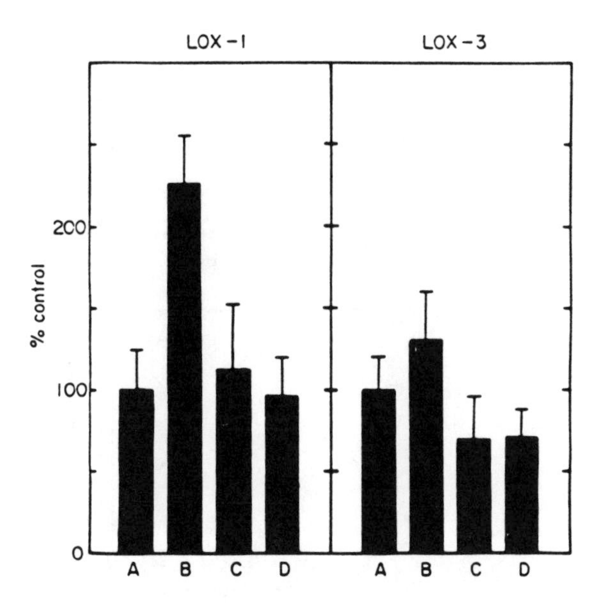

Fig. 67. Lipoxygenase-1 and -3 activities in leaves of *P. vulgaris* cv. Red Mexican 18 h after inoculation with: A. buffer alone: B, Race 1 of *P. s.* pv *phaseolicola;* C. Race 1 + 20 μg/ml cycloheximide; D, 20 μg/ml cycloheximide. Croft *et al.*, 1990.

decreased at 4 h post-inoculation and at 6 h after inoculation the activity was about the same as at 2 h after inoculation. It is hard to imagine that lipolytic acyl hydrolase plays an initial role in the HR induction when in French beans the induction of HR is completed 3 to 4 h after inoculation (Roebuck *et al.*, 1978). Indeed, in a subsequent study Koch *et al.* (1992) reported an induction of lipoxygenase mRNA in tomato by 3 h after inoculation with *P. s.* pv. *syringae* and an increased enzyme activity by 6 h with a maximum level by 24-48 h (Fig. 68). The timing of these changes clearly occurred after HR had been initiated. A dramatic increase in the activity of phosphatidase in tobacco, most likely related to the increased activity of lipolytic acyl hydrolases, was observed as early as 2 h after inoculation with an incompatible *E. amylovora* (Huang and Goodman, 1970). However, lipoxygenase activation in this host was not investigated.

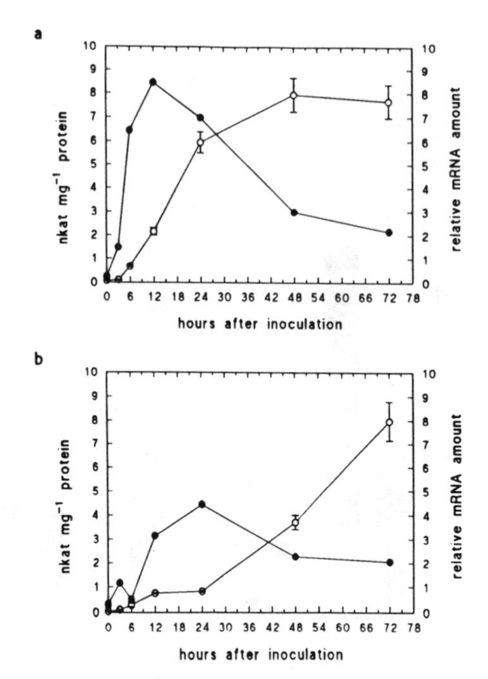

Fig. 68. Time course of changes in LOX exzyme activity (O) and relative amount of LOX mRNA (●) in leaves inoculated with (a) *P. s.* pv. *syringae* (HR) and (b) *P. s.* pv. *tomato* (susceptible reaction). Koch *et al.*, 1992.

Cellular control of oxygen free radicals: As mentioned before, active oxygen species in healthy tissue are strictly controlled by superoxide dismutase (SOD) and catalase. The activity of SOD during HR was found to be unaffected in cucumber/*P. s.* pv. *pisi* (an incompatible combination), cucumber/*P. s.* pv. *lachrymans* (a compatible combination) or cucumber/*P. fluorescens* (Keppler and Novacky, 1987) but in the French bean/*P. s.* pv. *phaseolicola* (incompatible) combination SOD activity increased 8 h after activation of lipoxygenase (Croft *et al.*, 1990). As discussed, treatments with SOD delayed, but did not prevent, the development of HR in cucumbers thus indicating that $O_2^{\cdot-}$ was involved in the observed HR development (Keppler and Novacky, 1987).

Oxidative burst: The plant response to fungal invasion (Doke *et al.*, 1987) and fungal elicitors (Apostol *et al.*, 1989a) is a sudden increase in active oxygen (for more details see page 64). Keppler *et al.* (1989) observed a similar transient oxidative burst in tobacco suspension cultured cells treated with bacteria: an HR-inducing strain of *P. s.* pv. *syringae*, an HR$^-$ mutant of *P. s.* pv. *syringae* and *P. s.* pv. *tabaci*, a tobacco pathogen. These authors found a dramatic transient burst of H_2O_2 between 0 and 1 h after exposing tobacco suspension culture cells to any of the three bacterial strains (Fig. 69). A transient burst of superoxide was also detected in all three tobacco cells/bacteria combinations (Fig. 70). The highest superoxide level was detected in the cells inoculated with the HR-inducing strain. The authors demonstrated that the decline following the active oxygen surge is a result of catalase activity. A second much smaller increase of H_2O_2 occurring between 3.5 and 4 h and of superoxide between 2 and 4 h after inoculation was more specific for HR-inducing bacteria.

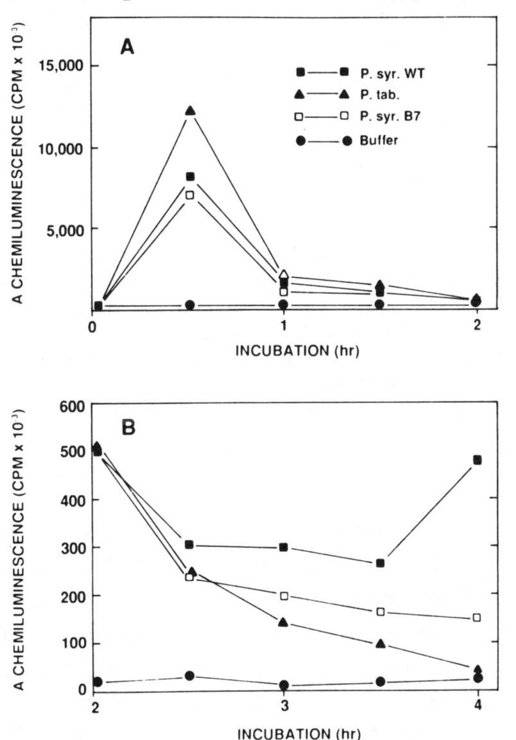

Fig. 69. Chemiluminescence (in the presence of luminol) in tobacco cell suspensions. Tobacco cell suspensions were equilibrated in assay medium for 1.5 hr and then exposed to bacteria (final concentration of 10^8 cfu/ml) in buffer. Bacteria used were: wild-type (*P. s.* pv. *syringae*, strain 61; induces a hypersensitive reaction); B7 (Tn5 insertion mutant of wild-type; does not induce a hypersensitive reaction); or *P. s.* pv. *tabaci* (compatible pathogen on tobacco). Keppler *et al.,* 1989.

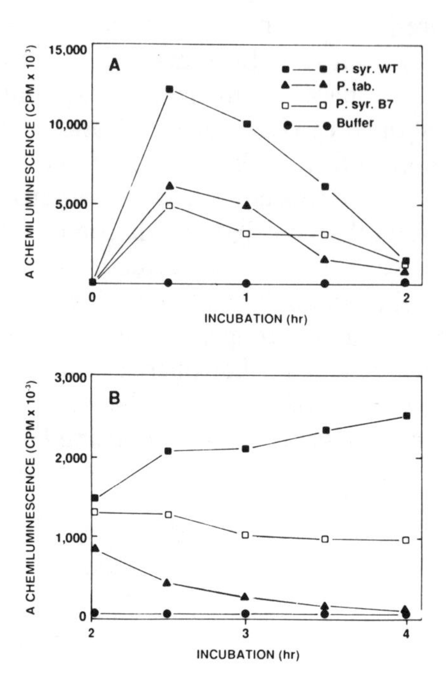

Fig. 70. Superoxide dismutase increased chemiluminescence (in the presence of luminol) in tobacco cell suspensions. Tobacco cell suspensions were equilibrated in assay medium for 1.5 hr and then exposed to bacteria (final concentration of 10^8 cfu/ml) in buffer, pH 5.6. Bacteria used were as described in preceding figure. Parallel 1.0-ml aliquots were transferred to scintillation vials. Superoxide dismutase (final concentration of 100 μg/ml) was added to one vial and incubated for 2 min prior to addition of luminol. Keppler *et al.*, 1989

Although the highest transient burst of active oxygen was detected in cells treated with the HR-inducing bacteria, the fact that a very similar increase in active oxygen occured also after treatment with bacteria that do not elicit HR indicates that the burst is a nonspecific event. Devlin and Gustine (1992) reported that an oxidative burst in *Trifolium repens* suspension cultured cells was elicited not only by the live HR-inducing bacteria, *P. corrugata*, but also with heat-killed bacteria, as well as treatment with mercury chloride (Fig. 71, 72). Thus in their experiments the character of the oxidative burst was again nonspecific. It is not clear what role this early nonspecific oxidative burst plays in later alterations that are more likely to be specific to HR-inducing bacteria, *e.g.* lipid peroxidation occuring 3 h after inoculation.

The non-specific oxidative burst may be related to a generalized induction of defense responses that are not correlated with the induction of HR *per se*. The separation of the two phenomena was demonstrated by Jakobek and Lindgren (1993) when transcripts for phenylalanine ammonia-lyase, chalcone synthase, chalcone isomerase and chitinase accumulated in common bean infiltrated with the wild-type *P. s.* pv. *tabaci* to a level similar to that in plants infiltrated with a *P. s.* pv. *tabaci* HR^- mutant, the non-pathogens *E. coli* and *P. fluorescens*, as well as a heat-killed HR^- strain. These experiments clearly demonstrate that HR is more specific and is not only part of the general defense response.

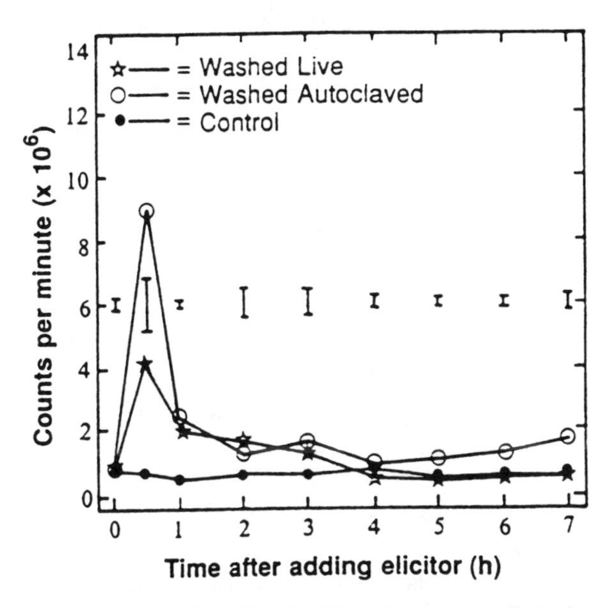

Fig. 71. Long-term elicitation of oxidative burst was studied using cells of live and autoclaved *P. corrugata* (0.4×10^8 cfu/ml) were added to *T. repens* plant suspension cultures. Chemiluminescence was measured at the indicated times during a 7-h period. Nonelicited, control plant cultures in buffer were observed in parallel with treated samples. Devlin and Gustine, 1992

The membrane lipid phase and/or a variety of other membrane sites like the K^+, Ca^{2+} channels, H^+- ATPase, and Ca^{2+}-ATPase (Freeman and Crapo, 1982; Fig. 73) may be structurally altered by oxygen free radicals. An example of such membrane changes may well be the alterations in tobacco chloroplast membrane proteins (decrease in mole percent nonpolar amino acids) 3 and 6 h after inoculation with *E. amylovora* that was reported by Huang and Goodman (1972).

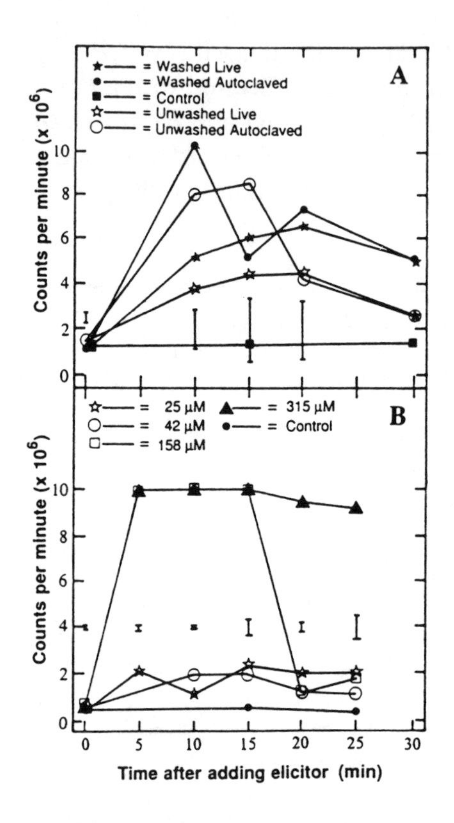

Fig. 72. Short-term elicitation of oxidative burst. Following addition of elicitors (A, *P. corrugata*; or B, $HgC1_2$) chemiluminescence was measured during a 30-min period. Experimental conditions and controls were otherwise the same as in the preceeding figure. Devlin and Gustine, 1992.

One animal tissue study clearly illustrates that even a brief exposure to free radicals results in immediate and significant damage to membrane channels in addition to lipid peroxidation (Mishra *et al.*, 1989). In this case, within one minute exposure, the activity of the Na^+/K^+ ATPase was decreased by 80% and membrane affinities for K^+ and ATP were altered, resulting in a deficit of energy for membrane transport.

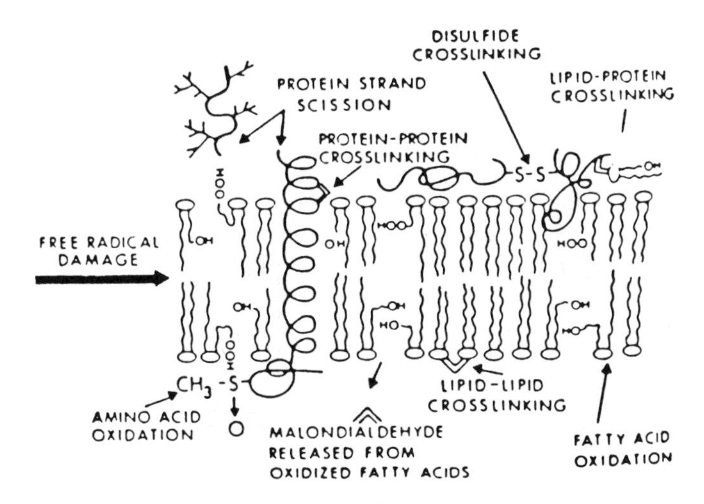

Fig. 73. Free radical damage to membranes. Free radicals can affect lipids by initiating peroxidation, which leads to short chain fatty acyl derivatives and the by-product malondialdehyde. A variety of cross-linking reactions can be mediated by malondialdehyde reactions. Free radicals can also catalyze amino acid oxidation, protein-protein cross-linking and protein strand scission. Freeman and Crapo, 1982.

The ion-channel defense model of Gabriel and Rolfe (1990) is worth mentioning at this point. According to this model, plant/bacteria recognition results from the binding of an *avr* gene product to the plasmalemma leading to the opening of ion channels. As a consequence of binding and subsequent events the host cell dies. Results from several studies of Atkinson and others may support such a model. However, lipid peroxidation elicited by HR-inducing bacteria suggests another possible model of "membrane opening" during HR. The interaction of *avr* or *hrp* gene product(s) *e.g.* harpin with membrane sites may result in the generation of active oxygen and cause structural damage with subsequent functional alteration (model of Apostol *et al.,* 1989a). In this case the first step in HR-elicitation is interaction of *avr* or *hrp* gene product(s) with some crucial component of membrane activity that leads to a cascade of membrane alterations and eventual HR cell death. How a *hrp* gene product(s) (*e.g.* harpin) affects the generation of oxygen radicals and its cellular control is still an open question. Since the oxidative burst itself is non-specific, harpin must continue to interact after this initial step. We suggest that continual production of harpin by a multiplying

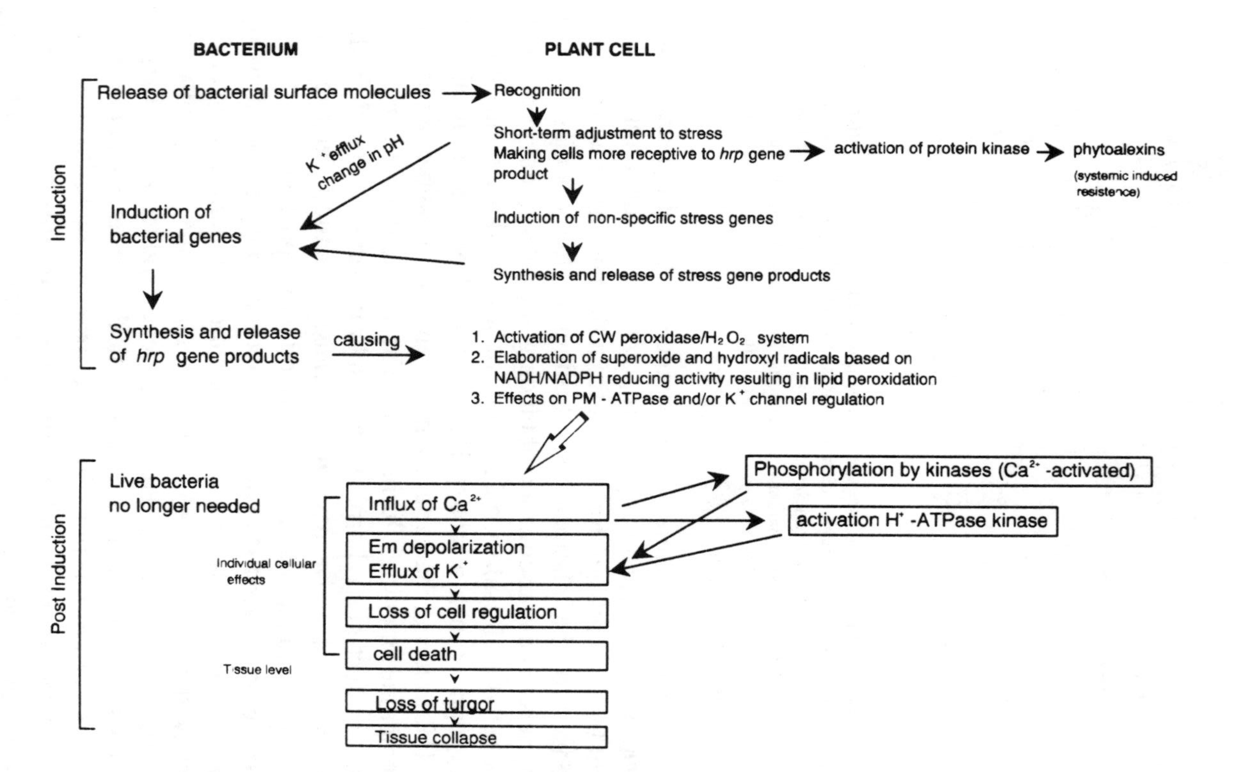

Fig. 74. Model of bacteria-induced HR. Novacky laboratory, unpublished.

bacterial population overwhelms the cellular antioxidant control system.

A possible sequence of events that may occur following recognition of the HR-inducing bacterium is presented in Fig. 74. This figure, the working model of Anton Novacky's laboratory, includes those interactions and events that have either direct support from the literature or appear to be reasonable extensions thereof.

Alterations of solute and water transport

The pressure probe developed by Zimmermann and Steudle (1975) and Hűsken *et al.* (1978) and the plasmometric method of Stadelmann (1969) were used to explore parameters of water transport during HR. With the pressure probe method three water transport parameters: the turgor pressure (P), the volumetric elastic modulus (ε) and the hydraulic conductivity (Lp) were monitored in single epidermal cells of *Nautilocalyx forgetii* (Gesneriaceae) leaves during HR elicited by the non-host (incompatible) pathogen *P. s.* pv. *pisi* (Pavlovkin and Brinckmann, 1989). With the plasmometric method the Lp along with the solute permeability and reflection coefficients were calculated for protoplasts of tobacco seedling stem parenchyma during HR induced by *P. s.* pv. *pisi* and during the compatible interaction with *P. s.* pv. *tabaci* (Turner, 1976; Turner and Novacky, 1976).

With both techniques dramatic alterations of Lp were found; however, the results were diametrically different and it is not clear what these alterations mean. The Lp determined with the pressure probe technique during HR in *Nautilocalyx* increased dramatically indicating increased water permeability of the plasma membrane. This Lp increase was found within the first hour after inoculation with *P. s.* pv. *pisi* and ranged from 19 to 42 x 10^{-8} m sec^{-1} x MP^{-1}. Within 8 h it had increased to 465 x 10^{-8} m sec^{-1} x MP^{-1}. In contrast the Lp values of tobacco stem parenchyma determined by the plasmometric method decreased 20 fold 2-3 h after inoculation.

Part of this difference lies in the components of the calculations and the influence of HR characteristics on these components. The P measured by pressure probe decreased within 2 h from 0.27 MPa to 0.11 MPa and was not measurable > 8 h post-inoculation. This alteration of turgor pressure, which probably reflects the loss of potassium (Stadelmann, personal communication), affects volume measurements and the calculation of ε and Lp. The plasmometric method which can be used near 0 MPa P would be less influenced by K^+ efflux. However, another reason for the difference may lie in the types of cells and tissues examined (Nonami and Boyer, 1990), *e.g.* epidermal *vs.* stem parenchyma, because measurements of the same type of cells with the two techniques have been shown to be very close (Stadelmann, personal communication). Clearly more studies are needed to determine how plant cell-membrane and wall components of water transport are affected by HR.

Status of stomata during HR

Wilting, a late stage of HR development, is accompanied by alterations in the diurnal regime of stomatal opening and closing. In daylight stomata are more closed in hypersensitively-responding leaves (Duniway, 1973) and cotton cotyledons (Popham *et al.*, 1993) than in healthy control tissue when measured with the porometer. Stomatal resistance of cotton cotyledons inoculated with *P. s.* pv. *tabaci* increased dramatically and became more variable on both upper and lower surfaces 6-8 h after inoculation (Fig. 75).

At night, partially open stomata were observed in hypersensitively responding cotton cotyledons, but not in the water-injected controls (Pike and Novacky, 1988). This occurred at the time when collapsed epidermal and mesophyll cells were observed and may simply be caused by loss of structural support as in epidermal strips (McRobbie, 1987).

In the final stage of HR, when tissues are drying, wide-open stomata border necrotic lesions (Pike and Novacky, 1988), terminating the transpiration stream at the edge of the lesion. This late wide opening could be a result of accumulated leaked K^+ or other plant products produced by the plant. Like many other characteristics of bacterial HR, it is not specific to bacteria-caused lesions, having been observed, for example, around virus-induced large necrotic lesions, (TRSV on lima bean, *Phaseolus lunatus*) (Pike and Sehgal, unpublished).

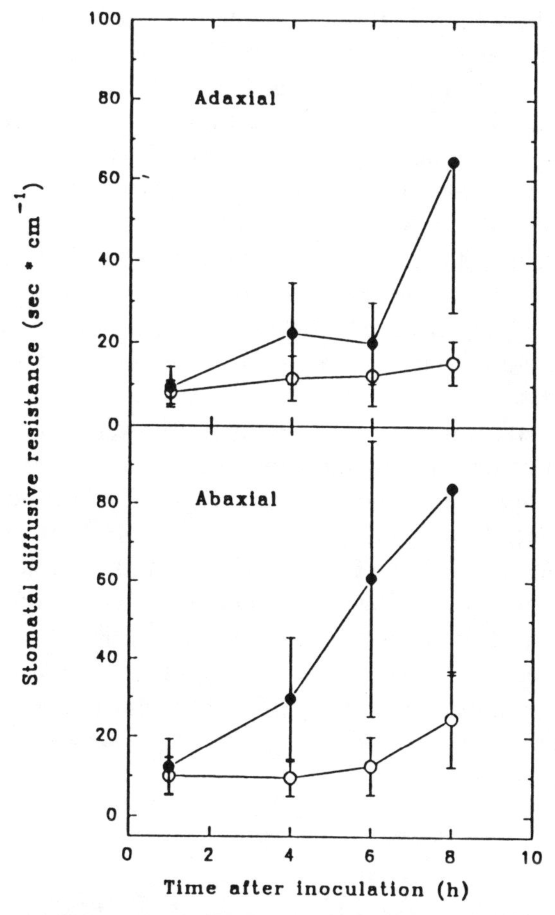

Fig. 75. The stomatal diffusive resistance (adaxial and abaxial) at various times after inoculation. (● - *P. s.* pv *tabaci*-treated, [HR] ○ - water-treated [control]) represent means calculated from between 11 and 16 observations, and error bars represent the standard deviation. Stomatal resistance measurements were significantly different (probability level 0.05) at 6 and 8 h after inoculation for both the adaxial and abaxial portions of the cotyledon. Popham *et al.*, 1993.

Conditions that alter the HR expression

The effect of scavengers of oxygen-free radicals on HR development is only one of several factors that can suppress bacterial HR. The induction and development of HR may be modified by different treatments or conditions that affect the plant response. Effects of such treatments are not unique to HR. We suggest that several treatments that affect the HR development may be attributed to an induced (acquired) plant resistance phenomenon (Sequeira, 1983). Lozano and Sequeira (1970b) prevented *P. solanacearum*-induced HR symptoms in tobacco by pretreatment with heat-killed bacterial cells of the same species. These authors followed the results of Lovrekovich and Farkas (1965) who had found that wildfire disease of tobacco (*P. s.* pv. *tabaci*) might be suppressed by pretreatment with heat-killed *P. s.* pv. *tabaci*. We suggest that in both situations we are dealing with the phenomenon of induced resistance (Sequeira, 1983), although, in one case it is the HR and in the other a disease that is prevented. The light-dependency of the protection and the spread of the protection to untreated areas suggest an active process.

The HR can be suppressed by both preinoculation with a low concentration of HR-inducing incompatible bacteria, and nonspecific injury. Preinoculation with 5 x 10^5 cells/ml of *P. s.* pv. *pisi* (a concentration that does not elicit visible HR) prevented the development of visible HR of tobacco and 10-fold drop in bacterial population following a challenge inoculation with the same bacteria at the concentration of 5 x10^6 cells/ml (Novacky *et al.*, 1973) (Fig. 76). Similarly, an injury caused by repeated water injection prevented HR symptom development in tobacco leaves subsequently inoculated with *P. s.* pv. *pisi* (Novacky and Hanchey,

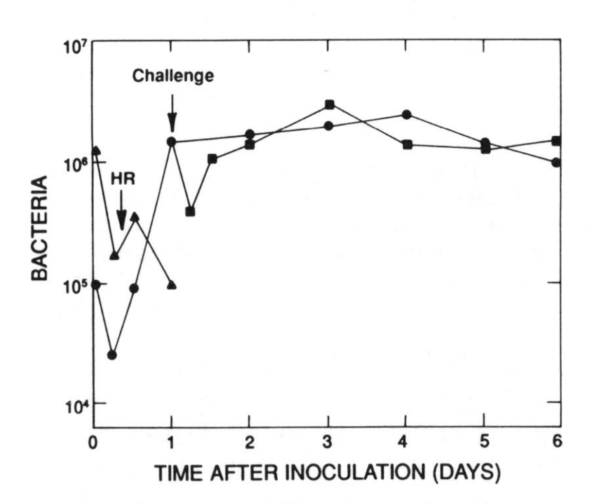

Fig. 76. Population of *P. s.* pv. *pisi* in the tobacco leaf after inoculation with 5 x 10 ^{5} cells/ml (●), 5 x 10^5 cells/ml challenged with 5 x 10^6 cells/ml (■) and 5 x 10^6 cells/ml (▲). Number of bacteria in the leaf determined in homogenates of three disks (1 cm in diameter) ground in 1 ml of H_2O. Novacky *et al.*, 1973.

1976). As with other effects of induced resistance (Sequeira, 1983), the protective effect was only transient and disappeared with time. A similar response of tobacco to water injection was reported by Godiard and coworkers (1990) who claimed that water, like HR-inducing bacteria, elicits the accumulation of pathogenicity-related proteins. The HR also can be prevented by ozone pretreatment (Pell *et al.*, 1977). Ozone injury, which is known to induce lipid peroxidation (Thompson *et al.*, 1987), would activate the free radical protective system, *i.e.* superoxide stimulation followed by scavenger enzyme activation. This might be the common denominator and basis of HR prevention by various injurious treatments.

Protection may also depend on the hormonal status of the plant tissue. The prevention of HR by cytokinin in tobacco leaves (Novacky, 1972; Haberlach *et al.*, 1978) and cucumber cotyledons (Pike and Novacky, unpublished) may be an illustration of such a case. Cytokinin may play a dual role in its interaction with the free radicals (Leshem, 1984). Thus, in addition to its activities as an "antisenescence" phytohormone, cytokinin may act as a direct scavenging agent of preexisting free radicals as reported by Würzburger *et al.* (1984). However, our experiments required cytokinin pretreatment 48 h prior to the inoculation with bacteria to suppress HR development. If cytokinins directly scavenged free radicals, simultaneous inoculation with bacteria and treatment with cytokinin should have prevented the development of HR. Efforts to repeat these experiments in cotton cotyledons were not successful (Novacky, unpublished) suggesting that hormonal status determining the "physiological age" of the plant may be important.

Treatments of cucumber cotyledons with salicylic acid another "plant hormone" (Raskin, 1992) also prevented HR after subsequent inoculation with 5 x 10^6 *P. s.* pv. *pisi* (Pike and Novacky, unpublished). This activity of salicylic acid is not unique to the bacteria-induced HR: in tobacco salicylic acid induced resistance to tobacco mosaic virus (White, 1979). Inoculation of cucumber leaves with *P. s.* pv. *syringae* resulted in both an increased level of salicylic acid in phloem sap 8-12 h after inoculation (Rasmussen *et al.*, 1991) and an induced resistance to *Colletotrichum lagenarium* (Smith *et al.*, 1991). The increase of salicylic acid in phloem sap supports the idea that this compound is transported to and induces resistance in other parts of the plant.

It is important to emphasize that HR is only prevented 12-24 h after the various pretreatments which strongly indicates that synthesis of plant products is necessary. This is illustrated well in experiments with salicylic acid in which other plant products such as peroxidases, pathogen-related proteins and phytoalexins have been reported (Raskin, 1992). However, there is no evidence that phytoalexins themselves prevent HR.

In all reported experiments of the prevention of HR-elicitation, only macroscopic symptoms were visually evaluated, hence we do not know what happened at the cellular level. It should be noted, that while confluent necrosis was prevented, *some wrinkling was present in our experiments indicating clearly that cell collapse occurred at the microscopic level.*

The same conditions that induce plant resistance may affect the bacterial side of the interaction. For example, several conditions have been shown to affect the bacterial HR-induction potential, the expression of *hrp* genes. Yucel *et al.* (1989) elegantly demonstrated that the conditions under which bacteria are cultured may influence the HR induction. They reduced the HR-induction period from 3h to 1h by coculturing *P. s.* pv. *pisi* or *P. s.* pv. *syringae* with tobacco suspension cells or by incubating these bacteria in a nitrogen-deficient medium which contained a metabolizable carbon source. The importance of nutritional status for the transcription of *hrp* genes and the expression of HR was also demonstrated for *Erwinia amylovora* (Wei *et al.*, 1992b). The *hrp* gene expression was regulated by medium composition as well as temperature (see page 179).

These observations may explain several older reports. Durbin and Klement (1977) observed HR suppression in tobacco by elevated temperature. They suggested at that time that 37°C affected some "temperature sensitive function" of the bacterium rather than of the plant. Hildebrand and Riddle (1971) explored the effect of light and temperature upon the HR induced in tobacco by various concentrations of pseudomonads, xanthomonads, erwinias and agrobacteria. Both factors caused rather dramatic effects which could result from bacterial gene induction. With the exception of *X. fragariae*, *Xanthomonas* spp.

induced HR under a narrower range of conditions than most of the pseudomonads tested. At low temperatures a reduced number of *P. s.* pv. *syringae*, *P. s.* pv. *phaseolicola* and *P. savastanoi* cells was required to induce visible HR, which can be interpreted as an increased HR-eliciting capacity (more plant cells killed per defined number of bacterial cells). The HR-induction at 22°C in soybean leaves inoculated with *P.s.* pv. *glycinea* race 1 and the development of pathogenesis in the same combination at elevated temperature (31°C) found by Keen and coworkers (1981) can be explained by the effect of temperature on *hrp* activation. However, such treatments may also affect the level of plant components that could activate the expression of *hrp* genes as, for example, the putative activating factor isolated by Schulte and Bonas (1992) (for details see page 178). In either case it would appear that HR is controlled or induced by temperature sensitive genes.

Genes Controlling Pathogenicity and the Hypersensitive Reaction

Historical aspects of *hrp* genes

Compatibility and incompatibility between bacteria and plants are controlled by a cluster of bacterial genes. These genes were first discovered when molecular biological techniques were introduced into the study of HR elicited by *P. s.* pv. *phaseolicola* (Anderson and Mills, 1985; Lindgren *et al.*, 1985) and pathogenicity caused by *P. s.* pv. *syringae* (Niepold *et al.*, 1985). Subsequently, Lindgren and coworkers (1986) identified a gene cluster encoding factors crucial to the expression of both basic compatibility and incompatiblity. Therefore, this gene cluster was named *hrp*, because both the HR ("*hr*") and pathogenicity ("*p*") are controlled by it.

It was shown that when bacteria lost the ability to induce HR on non-host plants (HR⁻) as a consequence of mutation by a transposon (Tn5) insertion they also lost their competence to colonize tissues in the susceptible host. Lindgren *et al.* (1986) identified seven mutations that were characterized by an HR⁻ phenotype. Of these, six had lost pathogenicity, *i.e.* the ability to form lesions associated with disease. Growth *in planta* of the seventh mutant, the only mutant strain that was able to colonize bean tissue, was reduced by an order of magnitude. The wild-type phenotype could be restored to all seven mutants by transformation with a single clone. Experiments revealed that *hrp* mutations were located in two sites of the *P. s.* pv. *phaseolicola* genome (Rahme *et al.*, 1991).

The *hrp* genes are not unique to *P. s.* pv. *phaseolicola*. Shortly after the discovery of the *hrp* gene cluster in *P. s.* pv. *phaseolicola* other researchers confirmed the existence of a *hrp* gene sequence in this pathovar, as well as in several other *P. syringae* pathovars and such distinctly different species as *P. solanacearum*, *Erwinia amylovora* and pathovars of *Xanthomonas campestris* (Willis *et al.*, 1991; Tab. 8). Whereas most of the data indicate that the interruption of the *hrp* gene segment results in the loss of pathogenesis, an exception has been also recorded. Azad and Kado (1984) reported that HR⁻ mutants of *Erwinia rubrifaciens* obtained by Tn*10* transposon insertion or nitrosoguanidine treatment remained pathogenic on their host, the English walnut *(Juglans regia)*.

Tab. 8. Studies that demonstrate hypersensitivity and pathogenicity (*hrp*) genes in plant pathogenic bacteria

P. s. pv. *phaseolicola*	(Anderson and Mills, 1985; Deasey and Matthyse, 1988; Somlyai *et al.*, 1986)
P. s. pv. *syringae*	(Anderson and Mills, 1985; Huang *et al.*, 1988; Niepold *et al.*, 1985),
P. s. pv. *tomato*	(Cuppels, 1986)
P. s. pv. *pisi*	(Malik *et al.*, 1987)
P. s. pv. *tabaci*	(Lindgren *et al.*, 1988)
P. s. pv. *glycinea*	(Huynh *et al.*, 1989)
P. solanacearum	(Boucher *et al.*, 1987; Huang *et al.*, 1990)
Xanthomonas campestris pathovars	(Daniels *et al.* 1988 ; Stall and Minsavage, 1990; Arlat *et al.*, 1991; Bonas *et al.*, 1991; Schulte and Bonas, 1992)
E. amylovora	(Bauer and Beer, 1987; Steinberger and Beer, 1988)

Several experiments have illustrated the role of *hrp* genes in HR induction. When a cosmid clone of *P. s.* pv. *syringae* 61 containing the *hrp* gene cluster with the *hrm* (regulatory) sequence was transferred to *P. fluorescens*, this saprophytic bacterium became HR-inducing (Huang *et al.*, 1988). A similar transfer conferred the HR-inducing ability into *E. coli* (Hutcheson *et al.*, 1989). All fourteen complementation groups within the *P. s.* pv. *syringae* genome were required for the expression of HR (Huang *et al.*, 1991). Either of the two bacteria non-pathogenic to plants, *P. fluorescens* and *E. coli*, elicited HR in several plant species from the *Solanaceae, Brassicaceae, Geraniaceae, Fabaceae, Cucurbitaceae* and *Poaceae* families when harboring the *hrp/hrm* gene cluster (Hutcheson *et al.*, 1993). Similarly, Beer *et al.* (1989) transferred a cosmid clone harboring the *hrp* gene of *E. amylovora* into *E. coli* and also conferred the HR-inducing ability.

Investigations are currently in progress to define the nature and function of the *hrp* gene product(s). Many researchers had attempted unsuccessfully to detect and isolate the elusive HR inducer since the discovery of bacterial HR (*e.g.* Sequeira and Ainslie, 1969; Klement, 1982; Hutcheson *et al.*, 1989). However, their failure appears now to have been not only the result of inadequate technology, but also an incomplete understanding of the HR concept *per se.* The identification of *hrp* genes in several plant pathogenic bacteria has enabled us not only to identify *hrp* gene product(s), the "HR elicitor molecule(s)", but also to search for signal molecules, *hrp* gene "activator molecules," originating in the host plant. Numerous recent *hrp* (and *avr*) gene studies have brought us closer to understanding the processes of HR activation and expression.

The identification of *hrp* (and *avr*) genes also permits examination of environmental factors that affect bacterial gene expression to promote or suppress HR. As previously discussed, the effect of these factors on the development of the bacterial HR were studied in the early days of bacterial HR research. We can expect that such findings will be elucidated by further study of *hrp* (and *avr*) gene regulation.

Control of *hrp* gene expression

Analysis of the regulatory protein encoded by *hrpS* of *P. s.* pv. *phaseolicola* demonstrated that plant and mammalian pathogens are probably regulated similarly. The *hrpS* is a 34.5 kD protein similar to several NtrC families of prokaryotic regulatory proteins of the enteric bacteria *Klebsiella pneumoniae*, *Escherichia coli* and *Salmonella typhimurinum* and the plant symbiotic bacteria *Rhizobium* spp. and *Bradyrhizobium japonicum*. This family of NtrC proteins functions as a "control system" by activating or repressing many different genes and operons. Grimm and Panopoulos (1989) proposed that the HrpS protein belongs to this group of proteins and thus *hrp* genes are under a general regulatory control. Rahme *et al.* (1991) also found similarity between the osmotic repression in their experiments and the osmotically-controlled virulence-associated genes in human pathogenic bacteria.

Further evidence of similarities between plant and mammalian pathogenic bacteria came from the U. Bonas and C. Boucher laboratories (Fenselau *et al.*, 1992; Gough *et al.*, 1992). Remarkable structural similarities between Hrp proteins from *X. c.* pv. *vesicatoria* and proteins from *Yersinia* (*Y. enterocolitica* and *Y. pestis*), *Shigella flexneri*, *Bacillus subtilis*, *Salmonella typhimurinum* and *Caulobacter crescentus* are suggestive of similar functions (Fenselau *et al.*, 1992). Observations of sequence similarity between Hrp proteins of *P. solanacearum* and virulence determinants of *Yersinia* species were reported also by Gough *et al.* (1992).

Export of *hrp* gene product(s)

The *hrp* gene products are apparently exported from cells via porins, channels across the outer bacterial membrane. Li *et al.* (1992) transferred a 25 kb *hrp/hrm* gene cluster of *P. s.* pv. *syringae* to *E. coli* to study phenotypic expression of the *hrp* gene cluster as judged by the ability to induce HR. Bacteria harboring the *hrp/hrm* cluster elicited HR on tobacco. Similar to the findings of Wingate *et al.* (1990), however, no immunoreactive surface features of these transformed *E. coli* were associated with the *P. s.* pv. *syringae hrp/hrm* cluster. *This suggested that surface expression of the hrp/hrm gene cluster was not important for mediating the HR response.* The authors further found that expression of the *hrp/hrm* gene cluster depended upon *omp*C and *omp*F, genes that code for the principal porins controlling the outer membrane permeability in *E. coli*. A mutant lacking these porins was unable to express HR. *These observations indicated that the HR-inducing factor(s), coded by the hrp/hrm cluster are secreted.* Since molecules no larger than 600 Daltons migrate through porins, Li *et al.*, (1992) suggest that the hrp/hrm HR-inducing product must be below 600 m.w.

***hrp* gene activation by the host plant or environmental factors**

Bacterial gene activation in response to host plant products has been reported in several plant-bacterial interactions. Several phenolic compounds with β-glycosidic linkage were reported to activate the *syr*B gene that controls syringomycin production in *P. s.* pv. *syringae* (Mo and Gross, 1991). Another example of bacterial gene activation by the host plant is the induction of *Rhizobium nod* genes by flavone, flavonone and isoflavone molecules (Long, 1989). Similarly, expression of the *virG* gene of *Agrobacterium tumefaciens* was shown to be stimulated by plant-phenolic compounds, like acetosyringone that are released upon wounding (Stachel *et al.*, 1985). Activation of *Agrobacterium tumefaciens* genes (*e. g., virG*) by phosphate starvation and acidic medium has been reported (Winans *et al.*,

1988). Similarly, Yucel *et al.* (1989) activated the HR-induction capacity of *P. s.* pv. *syringae* or pv. *pisi* by coculturing the bacteria with tobacco suspension culture cells prior to inoculating fresh tobacco cells. However, the activation resulted not only from coculturing bacteria with tobacco cells, but also from incubating bacteria in nitrogen-deficient media containing a metabolizable carbon source. The expression of HR, even in the presence of antibiotics, demonstrated *hrp* gene activation and indicated the presence of the *hrp* gene product.

Activation and expression of *hrp* genes in *X. c.* pv. *vesicatoria* in the host plant has been studied at the RNA level with RNA hybridization and β-glucuronidase gene fusion techniques (Schulte and Bonas, 1992). The six *hrp* genes A-F, were activated during growth in the plant tissue. Similar activation was found also after bacteria were treated with filtrates of tomato, pepper and tobacco suspension cultures cells (SCC). Of these, the tomato SCC was the most potent in promoting the β-glucuronidase-linked gene expression. Expression was found even after a 10-fold dilution of the tomato SCC filtrate. Additional experiments *led the authors to conclude that the factor responsible for hrp gene activation is a heat stable, hydrophilic organic molecule in the SCC filtrate that is < 1 kD in size.* It is important to note that *hrp* genes were activated not only with filtrates of the natural host (pepper), but also with those of the non-host (tobacco).

Wei *et al.* (1992b) elegantly illustrated the relationship between the pathogenicity of and hypersensitivity to *E. amylovara* by showing that the same genes function in both, but are expressed earlier during an incompatible association and appear later and to a lesser extent in the compatible situation. They followed *hrp* gene expression *in planta* using immature pear fruit host tissue and the non-host tobacco SCC. Inoculation of tobacco cells with *E. amylovara* induced expression of *hrp* genes 2 h after incubation with a maximum around 6 h. The *hrp* gene expression on pear fruits was delayed until 24 h and was only 10 to 20 % of the maximum activity in non-host tobacco cells. However, time of expression and percent activity could also have been affected by the very different conditions of bacterial growth on pear fruit *vs.* tobacco SCC.

Wei *et al.* (1992b) also used Tn*5* transposon mutagenesis with a β-glucuronidase reporter gene as a marker to test the influence of nutritional status on *hrp* gene expression in *E. amylovora*. Under rich medium conditions no HR expression was detected. Five *hrp* genes were well expressed in the plant tissue and after growth in an inducing medium containing mannitol, salts and $(NH_4)_2SO_4$. These genes were regulated by nitrogen and carbon sources, ammonium, nicotinic acid, pH of the medium and temperature. Low temperature was more favorable for *hrp* expression than high temperature.

Rahme *et al.* (1991) investigated the nature and role of environmental as well as plant signals in the activation of *hrp* genes. Results indicated that plant signals play a crucial role in inducing these genes in both compatible and incompatible associations. Different mechanisms apparently regulate individual loci of the *hrp* gene cluster *in planta* and *in vitro*; for example, the genes *hrpL* and *hrpRS* are expressed only in leaves, but not *in vitro*. Osmolarity and pH affected *hrp* loci differentially: a pH above 6.5 strongly diminished expression of *hrpAB, hrpC, hrpD,* and *hrpE,* whereas *hrpF* was not affected by pH at all. Likewise, the *hrpF* locus was unaffected by changes in the osmoticum, although the expression of *hrpAB, hrpC,* and *hrpD* was progressively reduced with increased concentrations of NaCl. *The findings of Rahme et al. (1991) demonstrate that the hrp gene cluster is under a complex regulatory system consisting of multiple physical factors, pH, osmotic strength and catabolite repression.* The fact that individual loci of the *hrp* cluster are apparently controlled differently *in vitro* and *in planta* strongly suggests that perhaps an array of plant signal molecule(s) plays an important role in *hrp* gene induction.

Harpin: the first bacterial elicitor

The *hrp* genes ushered in a new exciting phase in the search for gene products. An intensive process to identify these products followed the discovery that the *hrp* gene cluster is organized into several genetic complementation groups: *hrpA, hrpB, hrpC, hrpD,* and a regulatory *hrpS* in *P. s.* pv *phaseolicola* (Grimm and Panoupolos, 1989; Mindrinos *et al.*, 1990) or *hrpA* to *hrpF* in *X. c.* pv. *vesicatoria* (Bonas *et al.*, 1991). Identification of the direct gene products, *hrp*-encoded protein(s), however, has been only a first step towards understanding the elicitation of the HR. It was not known whether these proteins *per se* are the HR-elicitor(s) or whether HR-elicitation is a result of a cascade of events in which the *hrp* gene products play a role. It has become clear that further progress in the understanding of HR/plant resistance mechanisms requires the isolation of HR-eliciting molecules and the *in vitro* elicitation of HR.

In this regard, Wei *et al.* (1992a) have isolated a *hrp* product from *E. amylovora* and demonstrated its HR-eliciting activity both *in vitro* and *in planta*. No signs of host cell death were detected when they infiltrated the cell-free supernatants of either *E. amylovora* or *E. coli* harboring the *hrp* gene cluster of *E. amylovora* into plant tissue. However, a strong HR developed when they infiltrated centrifuged, filter-sterilized preparations from sonicated cells of *E. coli*. Purification revealed a 44 kD protein associated with the HR expression that the authors named **harpin.** Tobacco, tomato and *Arabidopsis thaliana* leaves were infiltrated with the protein at concentrations of >25 μg/ml. In all three species the HR developed within 12 h. Harpin elicited an early sign of HR, external alkalinization, in tobacco suspension cultured cells more rapidly than the bacteria themselves. The treatment of cells with harpin resulted in a typical pH increase within 1 h. It took 2-3 h for alkalinization to develop when tobacco cells were treated with *E. amylovora* bacterial cells.

The cell free HR-eliciting activity is heat stable and, not surprisingly, sensitive to proteases. The structural gene that codes for harpin consists of 1155 bases according to DNA sequencing analysis and has been labeled *hrpN*. It encodes a 385 amino acid peptide chain that is glycine-rich and has a high degree of hydrophilicity. Wei *et al.* (1992a) have demonstrated the location of harpin in the cell envelope. Antiserum raised in response to harpin reacted only with the membrane fraction of *E. amylovora.* A search in GenBank revealed a similarity with other glycine-rich proteins, mainly plant cell wall proteins and keratins.

This is the first reported induction of bacterial HR by an elicitor, the product of an identified gene, in the absence of live HR-eliciting bacteria. The discovery of harpin began a new phase in HR research. A search for harpins from other plant pathogenic bacteria followed this discovery. A similar protocol was employed by He *et al.* (1993) to identify the $harpin_{Pss}$ of *P. s.* pv. *syringae.*

The amino acid sequence of harpin$_{Pss}$, a 34.7 kD extracellular glycine-rich protein, is not similar to that of any known protein. It is not homologous to the *E. amylovora* harpin$_{Ea}$. However, there are similarities between the two harpins that indicate common structural features in HR-eliciting proteins. The gene encoding harpin$_{Pss}$ was identified as *hrpZ* and the gene controlling its secretion as *hrpH*. *Two directly repeated sequences in the carboxy-terminal 148 amino acid portion of harpin$_{Pss}$ were found sufficient to elicit the HR*. An interesting feature of harpin$_{Pss}$ is the absence of tyrosine which, as He *et al.* (1993) hypothesized, may be related to the passage of harpin$_{Pss}$ through the plant cell wall when H_2O_2-mediated cross-linking of tyrosine residues of cell wall proteins occurs.

The role of harpin in pathogensis is not clear yet. He *et al.* (1993) suggested that harpin likely functions in pathogenesis to release nutrients from plant cells to the apoplast. Hence both the identity of the harpin and the quantity produced would determine if HR or pathogenesis develops. However, these authors caution that other factors may modify the interaction. For example, animal pathogens like *Yersinia* spp. use a secretion pathway similar to the Hrp pathway but secrete multiple virulence proteins.

Avirulence genes and the gene-for-gene concept

In addition to *hrp* genes another class of genes, the avirulence (*avr*) genes, also express HR (Staskawicz *et al.*, 1984; 1987; Keen and Staskawicz 1988). These genes are more specific. They control the host cultivar range of physiological races, *i.e.* the phenotypic expression of HR (Mindrinos *et al.*, 1990; Keen 1990), resulting in virulence on some cultivars of plant species and avirulence on others. Experiments in which the *avrD* gene of *P. s.* pv. *tomato* was expressed in *E. coli* are good examples of this specificity (Keen *et al.*, 1990). This bacterium, nonpathogenic to plants, elicited the HR when carrying the *avrD* gene, but only in soybean cultivars that are incompatible with *P. s.* pv. *glycinea*. *Avr* genes have been identified in pseudomonads, xanthomonads as well as in *E. amylovora* (Tab. 9).

Tab. 9. Studies that first demonstrated avirulence (*avr*) genes in plant pathogenic bacteria

P. s. pv. *glycinea*	(Staskawicz *et al.*, 1984)
P. s. pv. *phaseolicola*	(Shintaku and Patil, 1986)
P. s. pv. *tomato*	(Kobayashi and Keen, 1986)
X. c. pv. *malvacearum*	(Gabriel *et al.*, 1986)
X. c. pv. *vesicatoria*	(Stall, 1986)
E. amylovora	(Vanneste *et al.*, 1987)

The *avr* genes were discovered in fungal pathogens when crosses between different physiological races were tested on plant cultivars carrying defined resistance genes (Flor, 1942). It was from Flor's studies that the gene-for-gene model emerged. According to this concept a single plant-resistance gene reacts with the matching single avirulence gene of a pathogen. The results of the interaction of these two genes is expressed as the HR (Keen, 1990). The site of this recognition, *i.e.* where the interaction between resistance gene product and the *avr* gene product takes place, has not been precisely established. However, a likely candidate is the plasma membrane, as our previous discussion has indicated.

The *avr* genes of plant pathogenic bacteria were discovered first in *P. s.* pv. *glycinea,* a pathogen of soybean (Staskawicz *et al.*, 1984). The first *avr* gene, *avrA,* was identified and sequenced by Napoli and Staskawicz (1987). Soybean cultivars that possess genes responsible for resistance against avirulent strains of *P. s.* pv. *glycinea* (*Rpg*) have been used in gene-for-gene studies (Keen, 1990). For example, *Rpg1,* a soybean gene described by Mukherjee and coworkers (1966), was found to complement the *avrB* gene of *P. s.* pv. *glycinea* (Staskawicz *et al.*, 1987). Avirulence genes from another pathovar, *P. s.* pv. *tomato*, also functioned as HR-eliciting genes on some soybean cultivars when cloned into *P. s.* pv. *glycinea*. The soybean gene *Rpg4* responds to bacteria harboring the *avr*D gene from *P. s.* pv. *tomato*, as well as HR elicitors produced by these bacteria (Keen *et al.*, 1990).

The activity of the *avr*D product, specific only to soybean cultivars resistant to *P. s.* pv. *glycinea* containing this gene, could be considered analogous to race and cultivar-specific elicitors detected in intracellular fluids of tomato infected with *Cladosporium fulvum* (de Wit and Spikman, 1982). Fungal host-selective toxins are also similar. For example, victorin C toxin produced by *Cochliobolus victoriae* (Wolpert and Macko, 1985) interacts with a toxin-binding protein in the cells of host tissue prior to the development of the toxic syndrome. This is analogous to an interaction of an *avr* gene product with the resistant gene product of an incompatible host. However, in the case of host-selective toxins it is the susceptible host that responds to the toxin produced by a virulent pathogen.

Gabriel *et al.* (1986) studied avirulence genes of *X. c.* pv. *malvacearum*. They found at least nine *avr* genes in this pathovar. Five of these genes specifically interacted with the individual resistance genes of compatible or susceptible lines of cotton. Interesting in their study is the finding of specificity in individual pairs of *avr* genes and their resistance genes.

In a subsequent investigation identifying six *avr* genes (De Feyeter and Gabriel, 1991), four appeared to interact in a gene-for-gene manner. An interesting observation in this study was the increased water soaking in the susceptible cultivar Acala 44 corresponding with the presence of the genes *avrb6* and *avrb7*. *Thus, avr genes may have a function in conditioning virulence of the pathogen on a susceptible host.*

Relationship between *hrp* and *avr* genes

Hrp and *avr* genes represent two distinct classes of genes. There is no evidence that *hrp* and *avr* genes are under control of some "common mechanism". However, *avr* genes may affect *hrp* gene expression or modify the activity of the *hrp* gene product (Hutcheson *et al.*, 1993). The *hrp* and *avr* genes have been investigated together in only a few host plant/bacteria combinations. In these the functional relationship between the two classes of genes is emerging. For example, *X. c.* pv. *vesicatoria* cells required the presence of *hrp* genes to express the *avrBs3* gene when inoculated into a pepper cultivar which carries the resistance locus *Bs3* (Knoop *et al.*, 1991). However, *X. c.* pv. *vesicatoria* grown in either minimal or complex medium expressed *avrBs3* independently of the *hrp* gene cluster.

It is important to note that while mutations of *hrp* genes result in a changed expression of HR and pathogenesis, mutations of *avr* genes alone do not affect pathogenic ability or infection efficiency (Gabriel and Rolfe, 1990). Therefore, *avr genes seem to be superimposed on the basic ability to parasitize and are even suggested to be a gratuitous phenomenon* (Gabriel, 1989).

***avr* gene products**

AvrD: The product of one *avr* gene, *avrD,* has been reported (Keen *et al.*, 1990). It was identified as a low molecular weight race-specific elicitor with a saturated alkyl chain and a gamma lactone ring. The biochemical function of the *avr*D gene product is at this writing unknown. However, a recent study of the molecular basis of symbiotic host specificity in *Rhizobium meliloti* (Roche *et al.*,1991) called attention to significant homology of the *Rhizobium* Nod H protein with the protein encoded by ORF4, a putative gene in the *avr*D operon of *P. s.* pv. *tomato* (Kobayashi *et al.*, 1990). This information not only confirms that a close genetic relationship exists between rhizobia and pseudomonads as indicated above for the *hrpS* and *ntr* genes, but also suggests avirulence in rhizobia might be controlled by *avr* genes.

AvrBs3: AvrBs3 may be a protein common to *X. campestris* pathovars. The expression of *avrBs3* from *X. c.* pv. *vesicatoria* was studied by Knoop *et al.* (1991) by fusion to β-glucuronidase. They found that *avrBs3* was expressed constitutively in both minimal and complex media as well as in the plant. The authors used polyclonal antibodies raised against a fusion protein in *E. coli* to identify a 122 kD protein in *X. c.* pv. *vesicatoria* cells. It is interesting to note that although the antibodies were specific for the AvrBs3 of *X.c.* pv. *vesicatoria*, they also recognized homologous proteins in other *Xanthomonas campestris* pathovars: *alfalfae, campestris, carotae, glycines, malvacearum* and *phaseoli*. The finding that AvrBs3 in *X. c.* pv. *vesicatoria* is located intracellularly in a soluble fraction will lead to futher studies to determine its export and interaction with the *Bs3* resistance gene.

An additional factor, however, must determine the specificity of individual races (Knoop *et al.*, 1991). In this respect attention is called to the high degree of similarity of the *avrBs3* gene to a gene identified by Swarup *et al*, (1992), the pathogenicity gene, *pthA* of *X. citri,* which is responsible for the Asiatic citrus canker disease. Apparently *pthA* is not only similar to *avrBs3,* but also is highly homologous to *X. c.* pv. *vesicatoria*, *avrBsP* and several *avr* genes of *X. c.* pv. *malvacearum*. When these authors transferred the *pthA* gene to *X. c.* pv. *malvacearum* and *X.c.* pv. *phaseoli* the resulting transconjugants remained nonpathogenic on citrus plants; however, they became incompatible with their respective hosts, *i.e.* in cotton and bean, and elicited HR in each case. On the other hand, the introduction of the *pthA* gene into mildly pathogenic or opportunistic *Xanthomonas* strains made them pathogens capable of causing citrus canker disease (Swarup *et al.*, 1991).

An important insight into the gene-for-gene relationship was gained from the analysis of the nucleotide sequence of the *avrBs3* gene from *X. c.* pv. *vesicatoria* (Herbers *et al.*, 1992). This gene encodes a 122 kD protein with 17.5 copies of a 34 amino acid repeat. The authors tried to determine if all of the almost identical repeat units are necessary for the HR-induction on the complementary pepper cultivar ECW-30R carrying the resistance gene *Bs3*. In addition, they explored the relationship between the repeats in this gene and race specificity. Plasmid derivatives with 3 to 17 repeats deleted were introduced into a *X. c.* pv. *vesicatoria* race 2 strain. Mutants were tested then on pepper cultivars ECW-30R containing the *Bs3* gene, nearly isogenic cultivar ECW and tomato cultivar Bonny Best. Only the mutant with a deletion of 3 repeats retained the capacity to elicit HR on ECW-30R. However, other mutants gained several new characteristics. For example, with the deletion of 4 repeating units, mutants aquired the ability to induce HR on ECW, the nearly isogenic cultivar of ECW-30R. Mutants of 6, 10 and 12 deletions induced a reaction intermediate between HR and water-soaked lesions on both ECW and ECW-30R. Herbers *et al.* (1992) found that a mutant with intermediate symptoms (mild HR and watersoaking) multiplied to a higher degree *in planta* than a mutant that produced clear HR symptoms. Most of the *avrBs3* derivatives elicited the HR in another plant species, tomato cv. Bonny Best, though the parental strain of *X. c.* pv. *vesicatoria* caused a susceptible reaction. These experiments, therefore, suggest that the deletion of repeats generated new characteristics of HR-elicitation.

The puzzling question for the authors was: "what distinctive feature of the *avrBs3* repetitive region gives rise to specificity?" Since the length of the repetitive region was not the basis for mutant specificity, the position of the repeat in the repetetive region must have been critical. The authors evaluated this question in more detail by selecting three mutants and determining their sequences. The sequences were identical to the original *avrBs3* except in the length of the repetitive region with the deletions occuring at different positions. One of these derivatives, designated Δ rep-9, with deleted repeats 5-7 did not induce HR on ECW-30R. On the other hand, the derivative Δ rep-109, with deleted repeats 13-15, induced the HR in this cultivar. *They concluded, therefore, that both the position of a repeat in the repetitive region and the presence or absence of certain repeats might produce differences of one or a few amino acids which then determine the avirulence.* No doubt these findings will be followed by detailed studies of the yet unknown plant genes complementary to avrBs3 that, according to these authors, may be widespread in the plant kingdom.

SUMMARY

We have assembled and discussed a comprehensive mass of literature concerning the hypersensitive defense reaction in plants. It is our purpose now to highlight the major controlling features of the phenomenon in order to define this general, very rapid reaction in plants and finally to call the attention of the reader to some unresolved questions that still exist in our understanding of HR. Our working hypothesis on how this phenomenon proceeds is based on specifically relevant data currently at hand and on correlative literature that appears germane to our view of the process *in planta*. Later in this summary we will present a detailed model of the communication between bacteria and plant cells as an example of the pathogen-host interaction that we believe ensues *generally* for HR.

There is little doubt now that HR is a resistance response in plants to viral, bacterial and fungal pathogens. The crucial characteristic of the phenomenon is the rapid host cell death that ensues following infection. The speed of the reaction sets HR clearly apart from the extended period of time required for typical disease symptoms to develop. Our definition of HR does not depart substantially from that reported by its discoverer, E. C. Stakman, in 1915, nor does it deviate conceptually from the description provided by Gäumann in 1950.

> **Hence, the hypersensitive reaction implies to us, *a priori*, rapid plant cell death following elicitation by a pathogen or metabolite thereof, reflecting electrolyte leakage due to membrane damage, the cessation of cyclosis, and cellular collapse. Attendant and subsequent to this sequence of events is the necrosis of the collapsed cells and localization of the eliciting pathogen. It is clear that the causative membrane damage occurs prior to and even in the absence of necrosis (Goodman, 1972 and Peever and Higgins, 1989)**

Communication between host cell and pathogen is a requisite event in the induction of HR. Evidence for this was revealed even in the early ultrastructural studies of HR (Goodman, 1972), in which the formation of vesicles and granular material was observed at the interface between responding plant cells and the pathogen. It seemed from these observations that HR was a highly focused "point reaction". More recently (Atkinson *et al.*, 1990, 1993), studies suggest complex plant signalling similar to that recognized in mammalian systems.

Fritig *et al.* (1987), revealed that TMV virus particles required a defined time of interaction with host cells in order to activate HR. Somlyai *et al.* (1991) reported similar data in a study of bacterial HR, thus proscribing that facet of HR characterized earlier by Klement (1982) as the induction phase of HR. It is plausible that this period of time is required for the expression of the genes (Wei *et al.*, 1992b) that are activated by the signals exchanged between bacteria and plant cells. Rahme *et al.*, (1991) found that plant signals are crucial in that individual loci of the *hrp* gene cluster are expressed differently *in planta* and *in vivo*. One such signal from the plant side has been isolated, a molecule that activates *hrp* genes in *X. c.* pv. *vesicatoria* (Schulte and Bonas, 1992). The fact that individual loci are controlled differently *in planta* and *in vitro* strongly suggests that an array of plant signal molecules functions in induction of the bacterial genes for HR and awaits identification. Once the induction period has been completed, living bacteria are no longer needed to sustain HR, as the phenomenon is no longer repressible with antibiotics (Klement and Goodman, 1967a). Apparently, once the *hrp* gene cluster has been expressed, complimentary plant reactions have been induced and HR must proceed to completion.

Precise data reveal that fungi must penetrate the cell wall before the phenomenon is activated (Tomiyama, 1956a). Both compatible and incompatible fungi grow at the same rate during the course of plant cell wall penetration. It is only after the plasma membrane is reached that metabolic, physiologic and morphologic differences become apparent. So it is with viruses, they must gain entry into the cell and they must replicate before HR is triggered. It is also apparent that virus *migration* to neighboring cells must take place. The latter is deduced from the data of Otsuki *et al.* (1972), who showed that at 22° C TMV inoculated into tobacco cvs. Samsun NN and Samsun, respectively, HR^+ and HR^-, spread of the virus occurs at approximately the same rate during the first 21 h of infection. The situation for bacteria is somewhat the same, although cell wall penetration does not occur. Both HR^+ and HR^- bacteria grow *in planta* at about the same rate for some 6-12 h, after which bacterial replication ceases when HR has been induced (Klement *et al.*, 1964)

There are published data revealing the precise nature of some of the elicitors of HR elaborated by viral, bacterial and fungal pathogens. However, in this summary we will focus on just three. Additionally, physiological, biochemical and ultrastructural evidence clearly reveals that the plasma membrane in particular, and membranes of subcellular organelles are the primary targets of elicitor molecules. There is also some evidence to suggest that the nuclear membrane is the most resistant to ultrastructurally observable alteration (Goodman and Plurad, 1971).

Elicitors of the fungus-induced HR have been prepared from fungal cell wall. A precise compound that elicits the action in potato by the model fungus *P. infestans*, is the fatty acid arachidonate (Bostock *et al.*, 1981). The gene controlling its synthesis has, as yet, not been proposed. However, the elicitor of HR in *Cf9* tomato genotype appears to be a peptide, a product of the *avr9* gene of *C. fulvum* (de Wit, 1992).

The specific elicitor produced by TMV is the virus coat protein, in those instances in which the host plant contains the N' gene (Dawson *et al., 1988).* It is now apparent that a mutation in a single nucleotide (cytosine-to-uracil) coding for a change in a single amino acid, phenylalanine instead of serine, is a controlling feature in the coat protein. Other single amino acid subsitutions are also known to elicit HR in N' tobacco. Recent data suggest that the nature of the inducer of HR by TMV in tobacco carrying the N gene, is a 126 kD ORF that codes for the protein inducer of N type resistance in Xanthi nc tobacco (Padgett and Beachy, 1993).

After a long and arduous search for a specific elicitor of HR by bacterial pathogens the elicitor elaborated by *E. amylovora* was elucidated in Steve Beer's laboratory (Wei *et al.*, 1992a, b). The synthesis of "harpin", a small heat stable protein that actually elicits HR is controlled by *hrp* genes. The protein is produced in both HR^+ and HR^- situations, however, in the former it is generated more rapidly and in greater amounts. Another biochemically different harpin has been isolated from *P. s.* pv. *syringae* (He *et al.*, 1993). We expect that other elicitors will be isolated because additional *hrp* genes have been reported in other bacteria (see Tab. 8). Structural similarities in *hrp* proteins from plant and mammalian pathogens (Fenselau *et al.*, 1992; Gough *et al.*, 1992), as well as similarities in a regulatory protein (Grimm and Panopoulos, 1989), suggest that additional similarities between plant and mammalian pathogens are yet to be revealed.

Each of the elicitors is resident at or near the surface of the inducing pathogen or is secreted by it. In case of the bacterial elicitor, harpin, secretion is essential (Li *et al.*, 1992). Some idea of the nature of the receptors for these molecules may be derived from the effect of the mutational rearrangement of coat protein of TMV. A shift in only one of the 158 amino acids in each of the 2200 subunits of the TMV coat protein causes both a structural change and a conformational alteration in its tertiary and quaternary configuration. It is believed that these modifications permit the elicitor to fit precisely the receptor's configuration on or near the plant cell surface. It is at that moment when contact occurs that the sequence of events that characterize HR begin.

HR Development

The rate at which HR begins to develop varies with each of the major groups of pathogens and and the plant surfaces with which they interact. *P. infestans* requires at least 70-90 min for penetration of the plant cell wall. Unlike the extended time required by viruses, bacteria need not replicate in order to exert their effect on a plant cell. There is good evidence that a single bacterial cell, in some combinations, can kill a single incompatible plant cell (Turner and Novacky, 1974). The magnitude of this reaction is placed in perspective when one compares the volumes of the interacting cells, 1 μm^3: 200,000 μm^3 (Goodman *et al.*, 1976). However, time is required for induction and secretion of bacterial *hrp* gene products.

Structural changes that develop in HR may be visualized with the aid of light and electron microscopy, as well as the naked eye. The initial visual change caused by *P. infestans* in potato leaf or tuber as seen with light microscopy is an increase in the rate of cyclosis which occurs in 10-60 min after wall penetration

(Tomiyama, 1956a,b). Then in the subsequent 90 min, cessation of cyclosis is followed by the onset of host cell discoloration, gelation of the cytoplasm and death of the first affected cells. Necrosis continues to intensify in these cells for an additional 10 h. The studies by Tomiyama (1956a,b), Bushnell (1982) and Meyer and Heath (1988a) all clearly link rapid cessation of cyclosis to the development of HR. This may occur in minutes after an HR-inducing fungus has breached the host cell wall. The inoculation of tobacco leaf tissue with TMV causes starch grains in chloroplasts to increase in size after 6 h (Weintraub and Ragetli, 1964). Early ultrastructural evidence revealed plasma membrane detachment from tobacco plant cell walls and chloroplast swelling 3 h after infiltration of leaf tissue with *P. s.* pv *pisi* (Goodman *et al.*, 1976).

Whether HR is induced by viruses, bacteria or fungi, induction is followed by electrolyte leakage. This increases rapidly and signifies severe plasma membrane damage. From electron micrographs we know that not only the plasma membrane sustains damage, but in addition, tonoplast, chloroplast and all other subcellular membranes as well. As a result of membrane perturbation, decompartmentalization occurs and numerous enzymes and their normally separated substrates interact. It is likely that necrogenesis reflects this condition.

We have found that not all cells in a leaf that has been infiltrated with HR-inducing bacteria will die, if the tissue is kept fully hydrated (Novacky *et al.*, 1993). This occurs in tissue that, when maintained under full aerobic conditions, would develop confluent cell collapse and necrosis. The basis for the survival of some of these cells is uncertain; however, failure of some plant cells to come in contact with bacteria is a reasonable assumption. Confluent necrosis develops when a threshold number of plant cells die (Turner and Novacky, 1974). The threshold depends upon temperature, nutrition, humidity and other environmental conditions as well as some specific characteristics of pathogen strain and cultivar of the plant species in question. All of these, no doubt, influence both bacterial gene induction (Yucel *et al.*, 1989; Wei *et al.*, 1992a; Schute and Bonas, 1992) and plant response .

Similarly, it was shown that the onset of necrosis associated with HR could be delayed by maintaining the induced tissue in a highly hydrated environment (Goodman, 1972). It was also apparent from this electronmicroscopical study that *although necrosis was delayed, membrane damage was not*. These observations imply the necessity of oxygen, and of water loss from membrane perturbed cells in order for confluent necrosis to develop. It has also been shown that a brief exposure to light is required for HR-related necrosis to develop (Peever and Higgins, 1989).

The primary environmental factor controlling HR, however, is temperature. It has long been known that that HR is controlled by temperature sensitive genes; specifically, this disease resistance response is maximally induced at 18-22°C and does not develop above 28-30°C. The apparent universality of the temperature sensitive gene in the induction of HR, suggests that it is present in all plants and that the complimentary *hrp*-like genes present in all HR-inducing plant pathogens are also temperature sensitive.

HR: a hypothetical model of the biochemical process

We turn now to a possible sequence of biochemical events that may be considered as participating in the development of HR. We are assuming that the fungal, viral and bacterial HR follow an identical pathway although it is possible that this is not the case. We know that the resistant-host HR *vs.* the non-host HR induced by bacteria proceeds more slowly and stems from a different genetic basis, avirulence *(avr)* genes. *Avr* genes, also the basis of fungal resistance, function specifically in relation to individual plant resistance genes. Several products of *avr* genes have been identified. They cause HR only in specific host(s) carrying the resistance genes. Unlike harpin which elicits HR in the non-host tobacco, the *avrD* product from *P. s.* pv. *tomato* only elicits HR in specific soybean cultivars (Keen *et al.*, 1991). Similarly the *avr9* product from *Cladosporium fulvum* (de Wit and Spikman, 1982) appears to be active only in those tomato plants carrying the *Cf9* gene. However, *hrp* genes were shown to be essential for *in planta* expression of *avrBs3* (Knoop *et al.*, 1991) which suggests that *avr* and *hrp* pathways to HR expression may converge.

In the preceding paragraphs of this summary we have presented physiological and structural evidence that the plasma membrane and other subcellular membranes are the primary targets of the elicitors. It is also apparent from the time differences in the induction of HR between tissues, isolated cells and particularly with protoplasts, in the case of *P. infestans*-potato interactions, that the plasma membrane is the site at which the cascade of events we call HR actually begins. Therefore, at this point we assume that the elicitor has reached the host cell plasma membrane, at least in the case of both bacterial and fungal HR-inducing pathogens. For the virus/host cell interactions wherein viral coat protein is the elicitor, we assume that sufficient coat protein has reached receptor sites. The following is our conception of the sequence of biochemical reactions that occur in the development of HR (Fig. 77).

It is not unreasonable to assume that in those instances in which the cell wall surrounds the plasma membrane, a second messenger is activated to transduce the wall impediment and perhaps the plasma membrane itself and that Ca^{2+} might fill this requirement. In a number of published studies concerning HR, an oxygen burst is detected within minutes or a few hours after cell suspension cultures are inoculated (*e.g.* Apostol *et al.*, 1989a). In several of these Ca^{2+} was integral to the development of the oxygen burst and the active oxygen species that was elaborated was H_2O_2; this was substantiated by the addition of catalase which could nearly totally suppress accumulation of the compound.

A study by Schwacke and Hager (1992) presented evidence that a protein-kinase was also an early player in the process. Its suppression by the specific inhibitor staurosporine provided good evidence for the probable participation of a phosphorylation event in the formation of an active oxygen species, a possibility that had been suggested much earlier by Tomiyama and his co-workers. Schwacke and Hager also suggested that the Ca^{2+} acts as a transducing element wherein protein kinase phosphorylates and thereby activates a cytosolic protein which in turn binds to membrane bound NADPH oxidase. Thus a reducing power source is established that can reduce molecular oxygen.

A calcium-dependent phosphorylation event has been linked to the oxidative burst caused by harpin, the bacterial elicitor of HR. The addition of $harpin_{Ea}$ (elicitor from *Erwinia amylovora*) to tobacco suspension cultured cells caused an active oxygen response within minutes (Baker *et al.*, 1993). K-252a, a protein kinase inhibitor, and $LaC1_3$, a calcium channel blocker, added immediately prior to $harpin_{Ea}$ eliminated this response indicating that a calcium-dependent phosphorylation is essential for the oxidative response. Similarly, recent studies in the Novacky laboratory (unpublished, 1993) have found that K-252a delayed the $harpin_{Ea}$-induced K^+/pH response and the membrane potential depolarization in tobacco suspension cultured cells, as well as the membrane depolarization in tobacco tissue segments.

The actual generation of active oxygen molecules from the reduced nucleotides, may be accomplished, according to Gross *et al.* (1977), as depicted in Fig. 24. The combined activities of membrane-bound malate dehydrogenase and wall-peroxidase (PER) support the formation of H_2O_2 with malate acting as the principal electron donor. The bound PER uses either NADH or NADPH as electron donors and produces both superoxide and H_2O_2. The superoxide may subsequently dismutate and form additional H_2O_2. However, with both superoxide and H_2O_2 being produced it is conceivable that in the presence of Fe^{2+} the highly reactive hydroxyl radical ($OH^{\cdot}$) may also be formed. It may in fact be the primary oxygen radical in causing intense membrane damage.

Although the deleterious effect of H_2O_2 on proteins has long been known, for example by oxidizing SH groups to S-S, its interaction with superoxide to form the hydroxyl radical has more recently brought these two oxygen species to the fore as participants in direct membrane destruction. The hydroxyl radical appears to be admirably suited to the role of HR-linked membrane damage and the localization of HR to the specific site in which the HR-inducer is injected. It appears that the elicitors of HR have little possibility for systemic movement and are therefore limited in their activities to sites neighboring their generation.

We should like to emphasize that the previously mentioned initial *oxidative burst* reported by several laboratories is a nonspecific event (Devlin and Gustine, 1992). The surge of H_2O_2 elaborated during the burst following the contact of plant and pathogenic bacterial cells occurs not only after contact with HR^+ strains but also with disease-causing as well as saprophytic bacteria. Even autoclaved bacterial cells or treatments with compounds like $HgCl_2$ foster this reaction.

The highly functional *endogenous cellular antioxidant system* strictly controls the level of active oxygen (Freeman and Crapo, 1982); thus the burst duration is rather short. The potassium efflux that accompanies the burst indicates that membrane perturbation occurs during the burst. However, cells apparently are not damaged irreversibly because leaked K^+ is later reabsorbed (W. Ullrich *et al.*, 1993). At this writing, the oxidative burst has been observed only in suspension cultured cells. There is no evidence that it occurs in inoculated tissues; hence it is difficult to compare the oxidative burst of suspension cultured cells with the increased active oxygen reported in hypersensitively responding tissues.

Although we presently do not know if *hrp*-gene products are present during the oxidative burst, ***we suggest that their continued output is responsible for the second, smaller, but sustained increase of active oxygen in suspension cultured cells (Keppler et al., 1989) as well as in tissues undergoing HR (Chai and Doke 1987).*** It is this second increase in active oxygen level that we believe is specific to the HR and disease-causing viruses, bacteria, and fungi. At this time lipoxygenase activity also increases contributing to free radical production.

We suggest that perturbation of the plasma membrane by the active oxygen species is responsible for the influx of Ca^{2+}. Several membrane components may be sites attacked by active oxygen (see Fig. 20) which permit the entry of Ca^{2+}. Low and Heinstein (1986) suggested that Ca^{2+} influx may accompany even the first oxidative burst in suspension culture cells. In some cases, *e.g.* saprophytic or autoclaved bacteria, the Ca^{2+} concentration is conceivably reduced by its adsorption, compartmentalization, or extrusion, and cell death due to the increased cytosolic Ca^{2+} is avoided. However, ultimately the increase in intracellular Ca^{2+} can no longer be prevented as has been observed when suspension cultured cells are treated with pathogenic, HR^+-inducing bacteria. The resultant sudden increase in intracellular Ca^{2+} induces K^+ efflux as documented by Atkinson *et al.* (1990).

A continuously elevated cytosolic Ca^{2+} concentration activates protein kinases and thus overwhelms the cell's coordinated control of structure and function. It is apparent that several enzymes maintain the structure and function of the plasma membrane, endoplasmic reticulum and mitochondria. The irreversible alteration of their regulation (by SOD, catalase, peroxidase, lipoxygenases etc.) most likely is responsible for the membrane damage and cell death that characterizes HR.

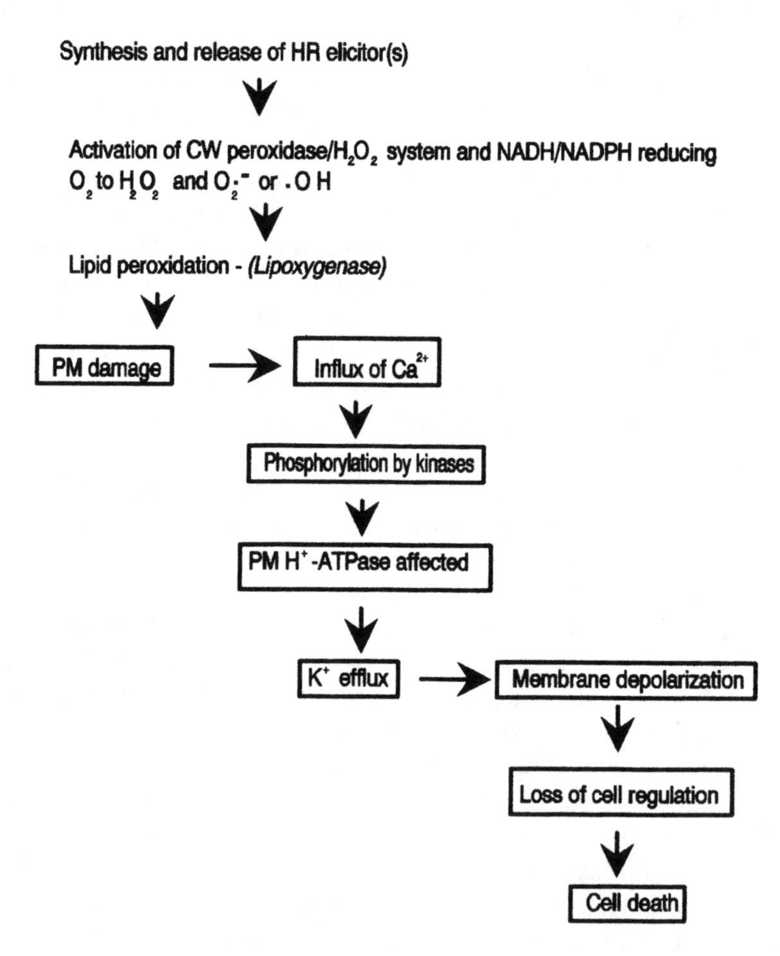

Fig. 77. Hypothetical sequence of events characterizing HR development

Some Unanswered Questions Pertaining To HR

1. Are there significant differences in HR induction and development between resistant host and non host plants?
2. Which of several active oxygen moieties are responsible for the induction of HR?
3. What is the role of Ca^{2+} in the induction of HR; is it in some way related to peroxidase H_2O_2 generation?
4. Is there a Ca^{2+}/calmodulin/ATPase interaction that activates peroxidase in the development of HR?
5. Do cyt *c* dependent NADPH oxidases play a feature role in HR development?
6. How do both acyl hydrolases and lipoxygenases contribute to HR-effected lipid metabolism?
7. What is the genetic control mechanism that governs the HR-eliciting activity of AA and EPA?
8. What role if any do glucans of pathogen origin play in the induction of HR set in motion by AA or EPA?
9. What is the threshold level of virus replication that induces HR?
10. Why are single protoplasts recalcitrant to HR development and expression in response to viral and bacterial pathogens?
11. What is the signal pathway for harpin activation of HR?
12. What is the relationship of HR to the initial nonspecific oxidative burst and K^+/H^+ response.
13. Are both the cell wall and the plasma membrane competent receptors for HR elicitors?
14. Do all HR elicitors activate the same pathway of HR development?
15. Are there essential modifiers or controllers of HR elicitors?
16. How do *hrp* and *avr* genes interact in bacterial elicitation of HR?
17. Can resistance genes responsible for HR be manipulated to produce increased resistance to specific diseases in plants?

Adam, A., Farkas, T., Somlyai, G., Hevesi. M., and Kiraly, Z. 1989. Consequence of O_2^- generation during a bacterially induced hypersensitive reaction in tobacco: deterioration of membrane lipids. Physiol. Mol. Plant Pathol. 34:13-26.

Allen, E.F., and Trewavas, A.J. 1987. The role of calcium in metabolic control. Pages 117-149 in: Biochemistry of Plants, Vol. 12. D.D. Davis, ed. Academic Press, NY.

Allen, F.H.E., and Friend, J. 1983. Resistance of potato tubers to infection by *Phytophthora infestans*: a structural study of haustorial encasement. Physiol. Plant Pathol. 22:285-292.

Allen, P.J., and Goddard, D.R. 1938. A respiratory study of powdery mildew of wheat. Am. J. Bot. 25:613-621.

Allen, R.F. 1923. A cytological study of infection of Baart and Kanred wheats by *Puccinia graminis tritici*. J. Agric. Res. 23:131-152.

Allen, R.F. 1926. Cytological studies of forms 9, 21 and 27 of *Puccinia graminis tritici* on Khapli Emmer. J. Agric. Res. 32:701-725.

Allington, W. B., and Chamberlain, D. W. 1949. Trends in the population of phytopathogenic bacteria within leaf tissue of susceptible and immune plant species. Phytopathology 39:656-660.

Al-Mousawi, A.H., Richardson, P.E., Essenberg, M., and Johnson, W.M. 1983. Specificity of the envelopment of bacteria and other particles in cotton cotyledons. Phytopathology 73: 484-489.

Anderson, D.M., and Mills, D. 1985. The use of transposon mutagenesis in the isolation of nutritional and virulence mutants in two pathovars of *Pseudomonas syringae*. Phytopathology 75: 104-108.

Apostol, I., Heinstein, P.F., and Low, P.S. 1989a. Rapid stimulation of an oxidative burst during elicitation of cultured plant cells. Plant Physiol. 90:109-116.

Apostol, I., Low, P.S., Heinstein, P., Sripanovic, R.D., and Altman, D.W. 1987. Inhibition of elicitor-induced phytoalexin formation in cotton and soybean cells by citrate. Plant Physiol. 84:1276-1280.

Apostol, I., Low, P.S., and Heinstein, P. 1989b. Effect of cell age of cell suspension cultures on susceptibility to a fungal elecitor. Plant Cell Rep. 7:692-695

Arlat, M., Gough, C.L., Barber, C.E., Boucher, C., and Daniels, M.J. 1991. *Xanthomonas campestris* contains a cluster of *hrp* genes related to the larger *hrp* cluster of *Pseudomonas solanacearum*. Mol. Plant-Microbe Interact. 4:593-601.

Arthur, J.C. 1929. The Plant Rusts (Uredinales), John Wiley and Sons, Inc., New York.

Askerlund, P., Larson, C., Widell, S., and Moller, I.M. 1987. NAD(P)H oxidase and peroxidase activities in purified plasma membranes from cauliflower inflorescences. Physiol. Plant. 71:9-19.

Atkinson, M.M. 1992. Signal transduction in the hypersensitive response of tobacco to *Pseudomonas syringae* pathovars. In: Biotechnology and Plant Protection. Bacterial Pathogenesis and Disease Resistance. D. Bills and S. D. Kung, eds. Butterworth Publishers. (In press)

Atkinson, M.M., and Baker, C.J. 1987a. Association of host plasma membrane K^+/H^+ exchange with multiplication of *Pseudomonas syringae* pv. *syringae* in *Phaseolus vulgaris*. Phytopathology 77:1273-1279.

Atkinson, M.M., and Baker, C.J. 1987b. Alteration of plasmalemma sucrose transport in *Phaseolus vulgaris* by *Pseudomonas syringae* pv. *syringae* and its association with K^+/H^+ exchange. Phytopathology 77:1573-1578.

Atkinson, M.M., and Baker, C.J. 1989. Role of the plasmalemma H^+-ATPase in *Pseudomonas syringae*-induced K^+/H^+ exchange in suspension-cultured tobacco cells. Plant Physiol. 91:298-303.

Atkinson, M.M., Bina, J., and Sequeira, L. 1993. Phosphoinositide breakdown during the K^+/H^+ exchange response of tobacco to *Pseudomonas syringae* pv. *syringae*. Mol. Plant-Microbe Interact. 6:253-260.

Atkinson, M.M., Huang, J.S., and Knopp, J.A. 1985a. Hypersensitivity of suspension-cultured tobacco cells to pathogenic bacteria. Phytopathology 75:1270-1274.

Atkinson, M.M., Huang, J.S., and Knopp, J.A. 1985b. The hypersensitive reaction of tobacco to *Pseudomonas syringae* pv. *pisi*. Activation of a plasmalemma K^+/H^+ exchange mechanism. Plant Physiol. 79:843-847.

Atkinson, M.M., Keppler, L.D., Orlandi, E.W., Baker, C.J., and Mischke, C.F. 1990. Involvement of plasma membrane calcium influx in bacterial induction of the K^+/H^+ and hypersensitive response in tobacco. Plant Physiol. 92:215-221.

Azad, H.R., and Kado, C.I. 1984. Relation of tobacco hypersensitivity to pathogenicity of *Erwinia rubrifaciens*. Phytopathology 74:61-64.

Babior, B.M. 1987. The respiratory burst oxidase. Trends Biochem. Sci. 12:241-243.

Bagi, G., and Farkas, G.L. 1968. A new aspect of the antistress effect of kinetin. Experientia 24:397-398.

Bailey, J.A. 1974. The relationship between symptom expression and phytoalexin concentration in hypocotyls of *Phaseolus vulgaris* infected with *Colletotrichum lindemuthianum*. Physiol. Plant Pathol. 4:477-488.

Bailey, J.A., and Deverall. B.J. 1971. Formation and activity of phaseollin in the interaction between bean hypocotyls (*Phaseolus vulgaris*) and physiological races of *Colletotrichum lindemuthianum*. Physiol. Plant Pathol. 1:435-449.

Bailey, J.A., and Ingham, J.L. 1971. Phaseollin accumulation in bean (*Phaseolus vulgaris*) in response to infection by tobacco necrosis virus and the rust *Uromyces appendiculatus*. Physiol. Plant Pathol. 1:451-456.

Bailey, J.A., and Rowell, P.M. 1980. Viability of *Colletotrichum lindemuthianum* in hypersensitive cells of *Phaseolus vulgaris*. Physiol. Plant Pathol. 17:341-345.

Bailey, J.A., Rowell, P.M., and Arnold, G.M. 1980. The temporal relationship between host cell death, phytoalexin accumulation and fungal inhibition during hypersensitive reactions of *Phaseolus vulgaris* to *Colletotrichum lindemuthianum*. Physiol. Plant Pathol. 17:329-339.

Baker, C.J., Atkinson, M.M., and Collmer, A. 1987. Concurrent loss in Tn5 mutants of *Pseudomonas syringae* pv. *syringae* of the ability to induce the hypersensitive response and host plasma membrane K^+/H^+ exchange in tobacco. Phytopathology 77:1268-1272.

Baker, C.J., Neilson, M.J., Sequeira, J., and Keegstra, K.G. 1984. Chemical characterization of the lipopolysaccharide of *Pseudomonas solanacearum*. Appl. Environ. Microbiol. 47: 1096-1100.

Baker, C.J., Orlandi, E.W., and Mock, N.M. 1993. Harpin, an elicitor of the hypersensitive response in tobacco caused by *Erwinia amylovora,* elicits active oxygen production in suspension cells. Plant Physiol. 102:1341-1343.

Bangerth, F. 1979. Calcium-related physiological disorders in plants. Annu. Rev. Phytopathol. 17:97-122.

Barber, M.S., Bertram, R.E., and Ride, J.P. 1989. Chitin oligosaccharides elicit lignification in wounded wheat leaves. Physiol. and Mol. Plant Pathol. 34:3-12.

Barnett, H.L. 1959. Plant disease resistance. Annu. Rev. Microbiol. 13:191-210.

Barton-Willis, P.A., Roberts, P.D. Gue, A., and Leach, J.E. 1989. Growth dynamics of *Xanthomonas campestris* pv. *oryzae* in leaves of rice. Phytopathology 79:573-578.

Barton-Willis, P.A., Wang, M.C., Staskawicz, B., and Keen, N.T. 1987. Structural studies on the O-chain polysaccharides of lipopolysaccharides from *Pseudomonas syringae* pv. *glycinea*. Physiol. and Mol. Plant Pathol. 30:187-197.

Bashan, Y., Ohon, Y., and Henis, Y. 1982. Detection of a necrosis-inducing factor of non-host plant leaves produced by *Pseudomonas syringae* pv. *tomato*. Can. J. Bot. 60:2453-2460.

Bauer, D.W., and Beer, S.V. 1987. Cloning of a gene from *Erwinia amylovora* involved in induction of hypersensitivity and pathogenicity. Pages 425-429 in: Plant Pathogenic Bacteria. E.L. Civerolo, A. Collmer, R.E. Davis, and A.G. Gillaspie, eds. M. Nijhoff, Boston.

Bayer, M.S. 1969. Gas chromatographic analysis of acidic indole auxins in *Nicotiana*. Plant Physiol. 44:267-271.

Beachy, R.N., and Murakishi, H.H. 1971. Local lesion formation in tobacco tissue culture. Phytopathology. 61:877-878

Beckman, C.H., Mueller, W.C. and Mace, M.E. 1974. The stabilization of artificial and natural cell wall membranes by phenolic infusion and its relation to wilt disease resistance. Phytopathology 64:1214-1220.

Beer, S.V., Zumoff, C.H., Bauer, D.W., Sneath, B.J., and Laby, R.J. 1989. The hypersensitive response is elicited by *Escherichia coli* containing a cluster of pathogenicity genes from *Erwinia amylovora* (Abstr.) Phytopathology 79:1156.

Beleid El-Moshaty, F.I., Pike, S.M., Novacky, A.J., and Seghal, O.P., 1993. Lipid peroxidation and active oxygen production in cowpea (*Vigna unguiculata*), leaves infected with tobacco ringspot virus or southern bean mosaic virus. Physiol. Plant Pathol. (in press)

Bell, J.N., Dixon, R.A., Bailey, J.A., Rowe, P.M., and Lamb, C.J. 1984. Differential induction of chalcone synthase mRNA activity at the onset of phytoalexin accumulation in compatible and incompatible plant-pathogen interactions. Proc. Natl. Acad. Sci. USA. 81:3384-3388.

Bell, J.N., Ryder, T.B., Wingate, V.P.M., Bailey, J.A., and Lamb, C.J. 1986. Differential accumulation of plant defense gene transcripts in a compatible and an incompatible plant-pathogen interaction. Mol. Cell. Biol. 6:1615-1623.

Bertl, A., Felle, H., Bentrup, F.-W. 1984. Amine transport in *Riccia fluitans*. Cytoplasmic and vacuolar pH recorded by a pH-sensitive microelectrode. Plant Physiol. 76:75-78.

Bonas, U., Schulte, R., Fenselau, S., Minsavage, G.V., Staskawicz, B.J., and Stall, R. E. 1991. Isolation of gene cluster from *Xanthomonas campestris* pv. *vesicatoria* that determines pathogenicity and the hypersensitive response on pepper and tomato. Mol. Plant Microbe Interact. 4:81-88.

Borchert, R., 1978. Time course and spatial distribution of phenylalanine ammonia-lyase and peroxidase activity in wounded potato tuber tissue. Plant Pathol. 62:789-793.

Bostock, R.M., Kuc, J.A., and Laine, R.A. 1981. Eicosapentaenoic and arachidonic acids from *Phtytophthora infestans* elicit fungitoxic sesquiterpenes in potato. Science 212:67-69.

Bostock, R.M., Laine, R.A., and Kuc, J.A. 1982. Factors affecting the elicitation of sesquiterpenoid phytoalexin accumulation by eicosapentaenoic and arachidonic acids in potato. Plant Physiol. 70:1417-1424.

Bostock, R.M., Nuckles, B., Henfling, F.W.D.M., and Kuc, J.A. 1983. Effects of potato tuber age and storage on sesquiterpenoid stress metabolite accumulation, steroid glycoalkaloid accumulation, and response to abscisic and arachidonic acids. Phytopathology 73:435-438.

Bostock, R.M., and Stermer, B.A. 1989. Perspectives on wound healing in resistance to pathogens. Annu. Rev. Phytopathol. 27:343-71.

Boucher, C.A., Barberis, P.A., and Arlat, M. 1988. Acridine orange selects for deletion of *hrp* genes in all races of *Pseudomonas solanacearum*. Mol. Plant-Microbe Interact. 1:282-288.

Boucher, C.A., Van Gijsegem, F., Barberis, P.A., Arlat, M., and Zischek, C. 1987. *Pseudomonas solanacearum* genes controlling both pathogenicity on tomato and hypersensitivity on tobacco are clustered. J. Bacteriol. 169:5626-5632.

Bricage, P. 1986. Isoperoxidases, markers of surrounding and physiological changes, *in situ* in leaves and *in vitro* in calli of *Pedilanthus tithymaloides* L. *Variegatus, Euphorbiaceae*: cell compartmentation and polyfunctionality, control of activity by phenols and specific roles. Pages 261-265 in: Molecular and Physiological Aspects of Plant Peroxidases. H. Greppin, C. Penel, and T. Gaspar, eds. Univ. of Geneva, Switzerland.

Briskin, D. P. 1990. Ca^{2+}-translocating ATPase of the plant plasma membrane. Plant Physiol. 94:397-400.

Brown, I., Mansfield, J., Irlam, I., Conrads-Strauch, J., and Bonas, U. 1993. Ultrastructure of interactions between *Xanthomonas campestris* pv. *vesicatoria* and pepper, including immunocytochemical localization of extracellular polysaccharides and the AvrBs3 protein. Mol. Plant-Microbe Interact. 6:376-386.

Brown, J.F., Shipton W.A., and White, N.H. 1966. The relationship between hypersensitive tissue and resistance in wheat seedlings infected with *Puccinia gramminis tritici*. Ann. Appl. Biol. 53:279-290.

Brown, O.R., and Seither, R.L. 1989. Paraquat inhibits NAD biosynthesis at the quinolinic acid synthetase site. Med. Sci. Res. 17:819-820.

Brown, O.R., and Seither, R.L. 1990. Paraquat toxicity and pyridine nucleotide coenzyme synthesis: a data correction. Free Radical Biol. & Med. 8:113-116.

Burden, R.S., Bailey, J.A., and Vincent, G.G. 1975. Glutinosone, a new antifungal sesquiterpene from *Nicotiana glutinosa* infected with tobacco mosaic virus. Phytochemistry 14:221-223.

Burkowicz, A., and Goodman, R.N. 1969. Permeability alterations induced in apple leaves by virulent and avirulent strains of *Erwinia amylovora*. Phytopathology 59:314-318.

Burr, T.J., and Katz, B.H. 1982. Isolation of *Agrobacterium tumefaciens* Biovar 3 from grapevine galls and sap, and from vineyard soil. Phytopathology 73:163-165.

Bushnell, W.R. 1981. Incompatibility conditioned by the *Mla* Gene, in powdery mildew of barley: The halt in cytoplasmic streaming. Phytopathology 71:1062-1066.

Cahill, D.M., and Ward, E.W.B. 1989. Rapid localized changes in abscisic acid concentrations in soybean in interactions with *Phytophthora megasperma* f. sp. *glycinea* or after treatment with elicitors. Physiol. Mol. Plant Pathol. 35:483-493.

Calder, V.L., and Palukaitis, P. 1992. Nucleotide sequence analysis of the movement genes of resistance breaking strains of tobacco mosaic virus. J. Gen. Virol. 73:165-168.

Cangelosi, G.A., Hung, L., Puvanesarajah, V., Stacey, G., Ozga, D.A., Leigh, J.A., and Nester, E.W. 1987. Common loci for *Agrobacterium tumefaciens* and *Rhizobium meliloti* exopolysaccharide synthesis and their roles in plant interactions. J. Bacteriol. 169:2086-2091.

Carpita, N., Sabrelarse, D., Monteginos, D., and Delmer, D. 1979. Determination of pore size of cell walls of living plants. Science 205:1144-1147.

Cason, E.T., Richardson, P.E., Essenberg, M.K., Brinkerhoff, L.A., Johnson, W.M., and Venere, R.J. 1978. Ultrastructural cell wall alterations in immune cotton leaves inoculated with *Xanthomonas malvacearum*. Phytopathology 68:1015-1021.

Castillo, F.J. 1986. Extracellular peroxidases as markers of stress? Pages 419-426 in: Molecular and Physiological Aspects of Plant Peroxidases. H. Greppin, C. Penel and T. Gaspar, eds. Univ. of Geneva, Switzerland.

Castillo, F.J., Penel, C., and Greppin, H. 1984. Peroxidase release induced by ozone in *Sedum album* leaves. Plant Physiol. 74:846-851.

Chai, H.B., and Doke, N. 1987. Superoxide anion generation: a response of potato leaves to infection with *Phytophthora infestans*. Phytopathology 77:645-649.

Chia, L.S., Thompson, J.E., and Dumbroff, E.B. 1981. Simulation of the effects of leaf senescence on membranes by treatment with paraquat. Plant Physiol. 67:415-420.

Chen, C.Y., and Heath, M. 1990. Cultivar specific induction of necrosis by exudates from basidiospore germlings of the cowpea rust fungus. Physiol. Mol. Plant Pathol. 37:169-177.

Clarke, A.E., Abbott, A., Mandel, T.E., and Pettitt, J. 1980. Organization of the wall layers of the stigmatic papillal of *Gladiolus gandavensis*: A freeze fracture study. J. Ultrastruct. Res. 73:269-281.

Clarke, A.E., and Gleeson, P.A. 1981. Molecular aspects of recognition and response in the pollen-stigma interaction, Pages161-211 in: Recent Advances in Phytochemistry Vol. 15, the Phytochemistry of Cell Recognition and Cell Surface Interactions. F.A. Loewus and C.A. Ryan, eds. Plenum Press, New York.

Coffey, M.D., Palevitz, B.A., and Allen, P.J. 1972. Ultrastructural changes in rust-infected tissues of flax and sunflower. Can. J. Bot. 50:1485-1492.

Collinge, D.B., and Slusarenko, A.J. 1987. Plant gene expression in response to pathogens. Plant Mol. Biol. 9:389-410.

Cook, A.A., and Stall, R.E. 1968. Effect of *Xanthomonas vesicatoria* on loss of electrolytes from leaves of *Capsicum annuum*. Phytopathology 58:617-619.

Cook, A.A., and Stall, R.E. 1969. Necrosis in leaves induced by volatile materials produced in vitro by bacteria. Phytopathology 59:259-260.

Cook, A.A., and Stall, R.E. 1971. Calcium suppression of electrolyte loss from pepper leaves inoculated with *Xanthomonas vesicatoria*. Phytopathology 61:484-487.

Cook, A.A., and Stall, R.E. 1977. Effects of watersoaking on response to *Xanthomonas vesicatoria* in pepper leaves. Phytopathology 67:1101-1103.

Corey, E.J., and Lansbury, P.T. Jr. 1983. Stereochemical course of 5-lipoxygenation of arachidonate by rat basophil leukemic cell (RBL-1) and potato enzymes. J. Am. Chem. Soc. 105:4093-4094

Creamer, J.R., and Bostock, R.M. 1988. Contribution of eicosapolyenoic fatty acids to the sesquiterpenoid phytoalexin elicitor activities of *Phytophthora infestans* spores. Physiol. Mol. Plant Pathol. 32:49-59.

Croft, K. P. C., Voisey, C. R., and Slusarenko, A. J. 1990. Mechanism of hypersensitive cell collapse: correlation of increased lipoxygenase activity with membrane damage in leaves of *Phaseolus vulgaris* (L.) inoculated with an avirulent race of *Pseudomonas syringae* pv. *phaseolicola*. Physiol. Mol. Plant Pathol. 36:49-62.

Crosse, J.E. 1965. Bacterial canker of stone-fruits, VI. Inhibition of leaf-scar infection of cherry by a saprophytic bacterium from the leaf surfaces. Ann. Appl. Biol. 56:149-160.

Culver, J.M., and Dawson, W.O. 1989a. Point mutations in the coat protein gene of tobacco mosaic virus induce hypersensitivity in *Nicotiana sylvestris*. Mol. Plant-Microbe Interact. 2:209-213.

Culver, J.M., and Dawson, W.O. 1989b. Tobacco mosaic virus coat protein: an elicitor of the hypersensitive reaction but not required for the development of mosaic symptoms in *Nicotiana sylvestris*. Virology 173:755-758.

Culver, J.M., and Dawson, W.O. 1991. Tobacco mosaic virus elicitor coat protein genes produce a hypersensitive phenotype in transgenic *Nicotiana sylvestris* plants. Mol. Plant-Microbe Interact. 4:458-463.

Culver, J.M., Lindbeck, A.G.C., and Dawson, W.O. 1991. Virus-host interactions: induction of chlorotic and necrotic responses in plants by tobamoviruses. Annu. Rev. Phytopathol. 29:193-217.

Cuppels, D.A. 1986. Generation and characterization of Tn*5* insertion mutations in *Pseudomonas syringae* pv. *tomato*. Appl. Environ. Microbiol. 51:323-327.

da Graca, J.V., and Martin, M.M. 1975. Ultrastructural changes in tobacco mosaic virus-induced local lesions in *Nicotiana tabacum* L. cv. "Samsun NN". Physiol. Plant Pathol. 7:287-291.

Daly, J. M. 1984. The role of recognition in plant disease. Annu. Rev. Phytopathol. 22:273-307.

Daniels, M.J., Dow, J.M., and Osbourn, A.E. 1988. Molecular genetics of pathogenicity in phytopathogenic bacteria. Annu. Rev. Phytopathol. 26:285-312.

Daub, M. E., and Hagedorn, D.J. 1980. Growth kinetics and interaction of *Pseudomonas syringae* with susceptible and resistant bean tissue. Phytopathology 70:429-436.

Davies, E. 1987. Action potentials as multifunctional signals in plants: a unifying hypothesis to explain apparently disparate wound responses. Plant, Cell Environ. 10:623-631.

Davis, K.R., Halbrock, K. 1987. Induction of defense responses in cultured parsley cells by plant cell wall fragments. Plant Physiol. 85:1286-1290.

Day, D.A., Aaron, G.P., and Laties, G. 1980. Nature and control of respiratory pathways in plants: Interaction of cyanide-resistant and cyanide sensitive pathway. Pages 197-241 in: Biochemistry of Plants, Vol. 2 P.K. Stumpf and E.E. Conn. eds., Academic Press.

Dawson, W.O., Bubrick, P., and Grantham, G.L. 1988. Modifications of the tobacco mosaic coat protein gene affecting replication, movement and symptomatology. Phytopathology 78:783-789.

Deasey, M.C., and Matthysse, A. G. 1988. Characterization, growth and scanning electron micrososcopy of mutants of *Pseudomonas syringae* pv. *phaseolicola* which fail to elicit a hypersensitive response in host and non-host plants. Physiol. Mol. Plant Pathol. 33:443-457.

De Feyter, R., and Gabriel, D.W. 1991. At least six avirulence genes are clustered on a 90-kilobase plasmid in *Xanthomonas campestris* pv. *malvacearum*. Mol. Plant-Microbe Interact. 4:423-432.

De Feyter, R., Yang, Y., and Gabriel, D. 1993. Gene-for-genes interactions between cotton *R* genes and *Xanthomonas campestris* pv. *malvacearum avr* genes. Mol. Plant-Microbe Interact. 6:225-237.

Deitrich, J.H. 1988. Role of abscisic acid in a Ca^{++} second messenger system. Physiol. Plant. 72:637-641

de Laat, A.M.M., and van Loon, L.C. 1983a. The relationship between stimulated ethylene production and symptom expression in virus-infected tobacco leaves. Physiol. Plant Pathol. 22:261-273.

de Laat, A.M.M., and van Loon, L.C. 1983b. Effects of temperature, light and leaf age on ethylene production and symptom expression in virus-infected tobacco leaves. Physiol. Plant Pathol. 22:275-283.

Devlin W.S., and Gustine, D.L. 1992. Involvement of the oxidative burst in phytoalexin accumulation and the hypersensitive reaction. Plant Physiol. 100:1189-1195.

de Wit, P.J.G.M. 1992. Molecular characterization of gene-for-gene systems in plant fungus interactions and the application of avirulence genes in control of plant pathogens. Annu. Rev. Phytopathol. 30:391-418.

de Wit, P.J.M.G., and Oliver, R.P. 1989. The interaction between *Cladosporium fulvum* and tomato: A model system in molecular plant pathology. Pages 227-236 in: Molecular Biology of Filamentous Fungi, Vol. 6. H. Nevalaimen, and N. Pentila, eds. Found. Biotechn. Industr. Ferment. Res.

de Wit, P.J.G.M., and Spikman, G. 1982. Evidence for the occurrence of race- and cultivar-specific elicitors of necrosis in intercellular fluids of compatible interactions between *Cladosporium fulvum* and tomato. Physiol. Plant Pathol. 21:1-11.

Dhindsa, R.S., Dhindsa, P.P., and Thorpe, T.A. 1981. Leaf senescence: correlated with increased levels of membrane permeability and lipid peroxidation and decreased levels of superoxide dismutase and catalase. J. Exp. Bot. 32:93-101.

Dhindsa, R.S., and Matowe, W. 1981. Drought tolerance in two mosses correlated with enzymatic defence against lipid peroxidation. J. Exp. Bot. 32:79-91.
Dietrich, A., Mayer, J.E. and Hahlbrock, K. 1990. Fungal elicitor triggers rapid, transient, and specific protein phosphorylation in parsley cell suspension cultures. J. Biol. Chem. 265:6360-6368.
Dixon, R.A., and Lamb, C.J. 1990. Molecular communication in interactions between plants and microbial pathogens. Annu. Rev. Plant Physiol. Plant Mol. Biol. 41:339-67.
Doke, N. 1975. Prevention of the hypersensitive reaction of potato cells to infection with an incompatible race of *Phytophthora infestans* by constituents of the zoospores. Physiol. Plant Pathol. 7:1-7.
Doke, N. 1983a. Involvement of superoxide anion generation in the hypersensitive response of potato tuber tissues to infection with an incompatible race of *Phytophthora infestans* and to the hyphal wall components. Physiol. Plant Pathol. 23:345-357.
Doke, N. 1983b. Generation of superoxide anion by potato tuber protoplasts during the hypersensitive response to hyphal wall components of *Phytophthora infestans* and specific inhibition of the reaction by suppressors of hypersensitivity. Physiol. Plant Pathol. 23:359-367.
Doke, N. 1985. NADPH-dependent O_2^- generation in membrane fractions isolated from wounded potato tubers inoculated with *Phytophthora infestans*. Physiol. Plant Pathol. 27:311-322.
Doke, N., and Chai, H.B. 1985. Activation of superoxide generation and enhancement of resistance against compatible races of *Phytophthora infestans* in potato plants treated with digitonin. Physiol. Plant Pathol. 27:323-334.
Doke, N., Chai, H.B., and Kawaguchi, A. 1987. Biochemical basis of triggering and suppression of hypersensitive cell response. Pages 235-251 in: Molecular Determinants of Plant Disease. S. Nishimura, ed. Springer-Verlag, Tokyo.
Doke, N., and Furuichi, N. 1982. Response of protoplasts to hyphal wall component in relationship to resistance of potato to *Phytophthora infestans*. Physiol. Plant Pathol. 21:23-30.
Doke, N., Garas, N.A., and Kuc, J. 1979a. Partial characterization and aspects of the mode of action of hypersensitivity-inhibiting factor (HIF) isolated from *Phytophthora infestans*. Physiol. Plant Pathol. 15:127-140.
Doke, N., Miura, Y., Chai, H-B., and Kawakita, K. 1991. Involvement of active oxygen in induction of plant defense response against infection and injury. Pages 84-96 in: Active Oxygen/Oxidative Stress and Plant Metabolism. E. Pell and K. Steffen, eds. Amer. Soc. Plant Physiol., Rockville, MD.
Doke, N., Nakae, Y., and Tomiyama, K. 1976. Effect of blasticidin S on the production of rishitin in potato tuber tissue infected by a compatible race of *Phytophthora infestans*. Physiol. Plant Pathol. 12:133-139.
Doke, N., and Ohashi, Y. 1988. Involvement of an O_2^- generating system in the induction of necrotic lesions on tobacco leaves infected with tobacco mosaic virus. Physiol. Mol. Plant Pathol. 32:163-175.
Doke, N., Sakai, S., and Tomiyama, K. 1979b. Hypersensitive reactivity of various host and nonhost plant leaves to cell wall components and soluble glucan isolated from *Phytophthora infestans*. Ann. Phytopathol. Soc. Jpn. 45:386-393.
Doke, N., and Tomiyama, K. 1975. Effect of blasticidin S on hypersensitive death of potato leaf petiole cells caused by infection with an incompatible race of *Phytophthora infestans*. Physiol. Plant Pathol. 6:169-175.
Doke, N., Tomiyama, K., Nishimura, N. and Lee, H.S. 1975. *In vitro* interactions between components of *Phytophthora infestans* zoospores and components of potato tissue. Ann. Phytopathol. Soc. Jpn. 41:425-433.

Doke, N., and Tomiyama, K. 1977. Effect of high molecular weight substances released from zoospores of *Phytophthora infestans* in the hypersensitive response of potato tubers. Phytopathol. Z. 90:236-242.

Doke, N., and Tomiyama, K. 1980a. Suppression of the hypersensitive response of potato tuber protoplasts to hyphal wall components by water soluble glucans isolated from *Phytophthora infestans*. Physiol. Plant Pathol. 16:177-186.

Doke, N., and Tomiyama, K. 1980b. Effects of hyphal wall components from *Phytophthora infestans* on protoplasts of potato tuber tissues. Physiol. Plant Pathol. 16:169-176.

Dong, X., Mindrinos, M., Davis, K.R., and Ausubel, F.M. 1991. Induction of *Arabidopsis* defense genes by virulent and avirulent *Pseudomonas syringae* strains and by a cloned avirulence gene. Plant Cell 3:61-72.

Drigues, P., Demery-Lafforgue, D., Trigalet, A., Dupin, P., Samain, D., and Asselineau, J. 1985. Comparative studies of lipopolysaccharide and exopolysaccharide from a virulent strain of *Pseudomonas solanacearum* and from three avirulent mutants. J. Bacteriol. 162:504-509.

Dunford, H.B. 1986. Catalytic mechanisms of plant peroxidases: with emphasis on reactions of compounds I and II. Pages 15-23 in: Molecular and Physiological Aspects of Plant Peroxidases. H. Greppin, C. Penel and T. Gaspar, eds. Univ. of Geneva, Switzerland.

Dunigan, D.D., Golemboski, D.B., and Zaitlin, M. Analysis of the *N* gene of *nicotiana*. Pages 120-135 in: Plant Resistance to Viruses, Vol. 133. Ciba Foundation Symposium. Wiley, Chichester.

Duniway, J.M. 1973. Pathogen-induced changes in host water relations. Phytopathology 63: 458-466.

Dunn, R.N., Hedden, P., and Bailey, J.A. 1990. A physiologically-induced resistance of *Phaseolus vulgaris* to a compatible race of *Colletotrichum lindemuthianum* is associated with increases in ABA content. Physiol. Mol. Plant Pathol. 36:339-349.

Durbin, R.D. and Klement, Z. 1977. High-temperature repression of plant hypersensitivity to bacteria: a proposed explanation. Pages 239-242 in: Proc. Symp. Hung. Akad. Sci. Z. Kiraly, ed. Akademiai Kiado, Budapest.

Dwurazna, M.M., and Weintraub, M. 1969a. Respiration of tobacco leaves infected with different strains of potato virus X. Can. J. Bot. 47:723-730.

Dwurazna, M.M., and Weintraub, M. 1969b. The respiratory pathways of tobacco leaves infected with potato virus X. Can. J. Bot. 47:731-736.

Dye, D.W., Bradbury, J.F., Goto, M., Hayward, A.C., Lelliott, R.A., and Schroth, M.N. 1980. International standards for naming pathovars of phytopathogenic bacteria and a list of pathovar names and pathotype strains. Rev. Plant Pathol. 59:153-168.

Ebel, J., Schmidt, W.P., and Lozal, R. 1984. Phytoalexin synthesis in soybean cells: Elicitor induction of phenylalanine ammonia-lyase and chalcone synthase mRNA's and correlation with phytoalexin synthesis accumulation. Arch. Biochem. Biophys. 232:240-248.

Ecker, J.R., and Davis, R.W. 1987. Plant defense genes are regulated by ethylene. Proc. Natl. Acad. Sci. USA. 84:5202-5206.

El-banoby, F.E., and Rudolph, K. 1979a. A polysaccharide from liquid cultures of *Pseudomonas phaseolicola* which specifically induces water-soaking in bean leaves (*Phaseolus vulgaris* L.) Phytopathol. Z. 95:38-50.

El-banoby, F.E., and Rudolph, K. 1979b. Induction of water-soaking in plant leaves by extracellular polysaccharides from phytopathogenic pseudomonads and xanthomonads. Physiol. Plant Pathol. 15:341-349.

El-Banoby, F.E., and Rudolph, K. 1981. The fate of the extracellular polysaccharide from *Pseudomonas phaseolicola* in leaves and leaf extracts from halo-blight susceptible and resistant bean plants (*Phaseolus vulgaris* L.). Physiol. Plant Pathol. 18:91-98.

El-Gewely, M.R., Smith, W.E., and Colotelo, N. 1972. The reaction of near-isogenic lines of flax to the rust fungus *Melampsora lini* I. Host-parasite interface. Can. J. Genet. Cytol. 14:743-751.

Ellingboe, A.H. 1972. Genetics and physiology of primary infection by *Erysiphe graminis*. Phytopathology 62:401-406.

Ellingboe, A.H. 1976. Genetics of host-parasite interactions. Pages 761-778 in: Encyclopedia of Plant Physiology. New Series, Vol 4. Physiological Plant Pathology. R. Herlefuss and P.H. Williams, eds. Springer Verlag Heidelberg, New York .

Elliott, D. C., and Skinner, J. D. 1985. Calcium dependent, phospholipid activated protein kinase in plants. Phytochemistry 25:39-44.

Elmhirst, J.F., and Heath, M.C. 1987. Interactions of the bean rust and cowpea rust fungi with species of the *Phaseolus* - *Vigna* plant complex. I. Fungal growth and development. Can. J. Bot. 65:1096-1107.

Elstner, E.F. 1982. Oxygen activation and oxygen toxicity. Annu. Rev. Plant Physiol. 33:73-96.

Elstner, E.F., and Heupel, A. 1976. Formation of hydrogen peroxide by isolated cell walls from horseradish (*Amoracia lapathifolia*.) Planta 130:175-180

Elstner, E.F., and Kramer, R. 1973. Role of the superoxide free radical ion in phytosynthetic ascorbate oxidation and ascorbate-mediated photophosphorylation. Biochim. Biophys. Acta. 314:340-353.

Elstner, E.F., Wagner, G.A., and Schutz, W. 1988. Activated oxygen in green plants in relation to stress situations. Plant Biochem. Physiol. 7:159-187.

Epperlein, M.M., Noronha-Dutra, A., and Strange, R.N. 1986. Involvement of the hydroxyl radical in the abiotic elicitation of phytoalexins in legumes. Physiol. Mol. Plant Pathol. 28:67-77.

Ercolani, G.L. 1970. Bacterial canker of tomato IV. the interaction between virulent and avirulent strains of *Corynebacterium michiganense* (E.F. Sm.) Jens. in vivo. Phytopathol. Mediterr. 9:151-159.

Ercolani, G.L. 1973. Two hypotheses on the aetiology of response of plants to phytopathogenic bacteria. J. Gen. Microbiol. 75:83-95.

Ercolani, G.L. 1984. Infectivity titration with bacterial plant pathogens. Annu. Rev. Phytopathol. 22:35-52.

Ercolani, G.L. 1985. The relation between dosage, bacterial growth and time for disease response during infection of bean leaves by *Pseudomonas syringae* pv. *phaseolicola*. J. App. Bacteriol. 58:63-75.

Ercolani, G.L., and Crosse. J.E. 1966. The growth of *Pseudomonas phaseolicola* and related plant pathogens *in vivo*. J. Gen. Microbiol. 45:429-439.

Ersek, T., Gaborjanyi, R., Holtzl, P., and Kiraly, Z. 1985. Sugar-specific attachment of *Pseudomonas syringae* pv. *glycinea* to isolated single leaf cells of resistant soybean cultivars. Phytopathol. Z. 113:260-270.

Esau, K., and Cronshaw, J. 1967. Relation of tobacco mosaic virus to the host cells. J. Cell Biol. 33:665-678.

Espelie, K.E., Franceschi, V.R., and Kolattukudy, P.E. 1986. Immunocytochemical localization and time course of appearance of an anionic peroxidase associated with suberization in wound-healing potato tuber tissue. Plant Physiol. 81:487-492.

Espelie, K.E., and Kolattukudy, P.E. 1985. Purification and characterization of an abscisic acid-inducible anionic peroxidase associated with suberization in potato (*Solanum tuberosum*). Arch. Biochem. Biophys. 240:539-545.

Essenberg, M., Cason, Jr., E. T., Hamilton, B., Brinkerhoff, L. A., Gholson, R. K., and Richardson, P.E. 1979. Single cell colonies of *Xanthomonas malvacearum* in susceptible and immune cotton leaves and the local resistant response to colonies in immune leaves. Physiol. Plant Pathol. 15:53-68.

Fagg, J., Woods-Tor, A., and Mansfield, J. 1991. Comparative study of interactions between *Bremia lactucae* and lettuce cells in suspension culture and in cotyledons. Physiol. Mol. Plant Pathol. 38:105-116.

Fahy, P.C., and Persley, G. J. 1983. Plant Bacterial Diseases: A Diagnostic Guide. Academic Press, London.

Farkas, G.L., Kiraly, Z., and Solymosy, P. 1960. Role of oxidative metabolism in the localization of plant viruses. Virology 12:408-421.

Favali, M.A., Conti, G.G., and Bassi, M. 1978. Modifications of the vascular bundle ultrastructure in the "Resistant Zone" around necrotic lesions induced by tobacco mosaic virus. Physiol. Plant Pathol. 13:247-251.

Feistner, G.J. 1988. (L)-2,5-Dihydrophenylalanine from the fireblight pathogen *Erwinia amylovora*. Phytochemistry 27:3417-3422.

Felix, G., Grosskopf, D.G., Regenass, M., Boller, T. 1991. Rapid changes of protein phosphorylation are involved in transduction of the elicitor signal in plant cells. Proc. Natl. Acad. Sci. USA 88: 8831-8834.

Fenselau, S., Balbo, I., and Bonas, U. 1992. Determinants of pathogenicity in *Xanthomonas campestris* pv. *vesicatoria* are related to proteins involved in secretion in bacterial pathogens of animals. Mol. Plant-Microbe Interact. 5:390-396.

Fernandez, M.R., and Heath, M.C., 1986. Cytological responses induced by five phytopathogenic fungi in a nonhost plant, *Phaseolus vulgaris*. Can. J. Bot. 64:648-657.

Fett, W.F., and Jones, S.B. 1982. Role of bacterial immobilization in race-specific resistance of soybean to *Pseudomonas syringae* pv. *glycinea*.. Phytopathol. 72:488-492.

Fett, W.F., Osman, S.F., Fishman, M.L., and Siebles, T.S. 1986. Alginate production by plant-pathogenic pseudomonads. App. Environ. Microbiol. 52:466-473.

Feys, M., Naesens, W., Tobback, P., and Maes, E. 1980. Lipoxygenase activity in apples in relation to storage and physiological disorders. Phytochemistry 19:1009-1011.

Fischer, E., and Novacky, A. 1965. Membrane potential, respiration, photosynthesis and ATP content in bell pepper fruit segments inoculated with *Xanthomonas vesicatoria*. (Abstr.) Plant Physiol. 65:107.

Fisher, D.A., and Bayer, D.E. 1972. Thin sections of plant cuticles demonstrating channels and wax platelets. Can. J. Bot. 50:1509-1511.

Flor, H.H. 1942. Inheritance of pathogenicity of *Melampsora lini*. Phytopathology 32: 653-669.

Fraser, R.S.S., and Loughlin, S.A.R. 1980. Resistance to tobacco mosaic virus in tomato: effects of the *Tm-1* gene on virus multiplication. J. Gen. Virol. 48:87-96.

Freeman, B.A., and Crapo, J.D. 1982. Biology of disease - free radicals and tissue injury. Lab. Invest. 47:412-426.

Frimer, A.A., Aljadeff, G., and Ziv, J. 1983. Reaction of (arylmethyl) amines with superoxide anion radical in aprotic media. Insight into cytokinin senescense inhibitor. J. Org. Chem. 48: 1700-1705.

Fritig, B., Kauffmann, S., Dumas, B., Geoffroy, P., Kopp, M., and Legrand, M. 1987. Mechanism of the hypersensitivity reactions of plants, Pages 91-108 in: Plant Resistance to Viruses, Wiley, Chichester.

Fry, P.R., and Mathews, R.E.F. 1963. Timing of early events following inoculation with tobacco mosaic virus. Virology 19:461-864.

Fry, S. 1981. Phenolic components of the primary wall and their possible role in hormonal regulation of growth. Planta 146:343-351.

Fry, S.C. 1986. Polymer-bound phenols as natural substrates of peroxidases. Pages 169-181 in: Molecular and Physiological Aspects of Plant Peroxidases. H. Greppin, C. Penel, and T. Gaspar, eds. Univ. of Geneva, Switzerland.

Funatsu, G., and Fraenkel-Conrst, H. 1964. Location of amino acid exchanges in chemically evoked mutants of tobacco mosaic virus. Biochemistry 3:1356-1361.

Furlong, C. E. 1987. Osmotic shock sensitive transport systems. Pages 768-796 in: *Escherichia coli* and *Salmonella typhimurinum*: Cellular and Moleuclar Biology. F.C. Neidhardt, ed. Amer. Soc. Microbiol. Washington, D.C.

Furner, I.J., Higgins, E.S., and Berrington, A.W. 1989. Single-stranded DNA transforms plant protoplasts. Mol. & Gen. Genet. 220:65-68.

Furuichi, N., and Suzuki, J. 1989. Isolation of proteins related to B-lectin from potato which bind hyphal wall compounds of *Phytophthora infestans*. J. Phytopathol. 127:281-290.

Furuichi, N., and Suzuki, J. 1990. Purification and properties of suppression glucan isolated from *Phytophthora infestans*. Ann. Phytopathol. Soc. Jpn. 56:457-467.

Furuichi, N., Tomiyama, K., and Doke, N. 1979a. Hypersensitive reactivity in potato: transition from inactive to active state induced by infection with an incompatible race of *Phytophthora infestans*. Phytopathology 69:734-736.

Furuichi, N., Tomiyama, K., Doke, N. and Nozue, M. 1979b. Inhibition of further development of hypersensitive reactivity to *Phytophthora infestans* by blasticidin S in cut tissue of potato tuber at various stages of aging process. Ann. Phytopathol. Soc. Jpn. 45:215-220.

Furuichi, N., Tomiyama, K., and Doke, N. 1980a. The role of potato lectin in the binding of germ tubes of *Phytophthora infestans* to potato cell membrane. Physiol. Plant Pathol. 16:249-256.

Furuichi, N., Tomiyama, K., and Doke, N. 1980b. Induction of hypersensitive reactivity in juvenile potato cell to compatible race of *Phytophthora infestans* by its infection *per se*. Ann. Phytopathol. Soc. Jpn. 46:247-249.

Furuichi, N., Tomiyama, K., and Doke, N. 1982. Effect of water soluble glucan of *Phytophthora infestans* on the agglutination of germinated cystospores caused by potato lectin. Ann. Phytopathol. Soc. Jpn. 48:234-237.

Gabriel, D.W. 1989. Genetics of plant parasite populations and host-parasite specificity. Pages 343-379 in: Plant-Microbe Interactions, Molecular and Genetic Perspectives, Vol. 3. T. Kosuge and E.W. Nester, eds. McGraw-Hill Publishing Co., New York.

Gabriel, D.W., Burges, A., Lazo, G.R. 1986. Gene-for-gene interactions of five cloned avirulence genes from *Xanthomonas campestris* pv. *malvacearum* with specific resistance genes in cotton. Proc. Natl. Acad. Sci. U.S.A. 83:6415-6419.

Gabriel, D.W., and Rolfe, B.K. 1990. Working models of specific recognition in plant-microbe interactions. Annu. Rev. Phytopathol. 28:365-391.

Garas, N.A., Doke, N., and Kuc, J. 1979. Suppression of the hypersensitive reaction in potato tubers by mycelial components from *Phytophthora infestans*. Physiol. Plant Pathol. 15:117-126.

Gardner, J. M., and Kado, C. I. 1972. Induction of the hypersensitive reaction in tobacco with specific high molecular weight substances derived from the osmotic shock fluid of *Erwinia rubrifaciens*. (Abstr.) Phytopathology 62:759.

Gaspar, T. 1986. Integrated relationships of biochemical and physiological peroxidase activities. Pages 455-470 in: Molecular and Physiological Aspects of Plant Peroxidases. H. Greppin, C. Penel, and T. Gaspar, eds. Univ. of Geneva, Switzerland.

Gaspar, T., Penel, C., Castillo, F.J., and Greppin, H. 1985. A two-step control of basic and acidic peroxidases and its significance for growth and development. Physiol. Plant. 64:418-423.

Gaumann, E. 1950. Principles of Plant Infection (Trans. by W.B. Breierley) Crosby Lockwood and Son, London.

Gianinazzi, S., Pratt, H.M., Shewry, P.R., and Miflin, B.J. 1977. Partial purification and preliminary characterization of soluble leaf proteins specific to virus infected tobacco plants. J. Gen. Virol. 34:345-351.

Giaquinta, R.T. 1983. Phloem loading of sucrose. Annu. Rev. Plant Physiol. 34:347-387.

Gilpatrick, J.D., and Weintraub, M. 1952. An unusual type of protection with the carnation mosaic virus. Science 115:701-702.

Godiard, L., Ragueh, F., Froissard, D., Leguay, J., Grosset, J., Chartier, Y., Meyer, Y., and Marco, Y. 1990. Analysis of the synthesis of several pathogenesis-related proteins in tobacco leaves infiltrated with water and with compatible and incompatible isolates of *Pseudomonas solanacearum*. Mol. Plant-Microbe Interact. 3:207-213.

Goff, C.W. 1975. A light and electron microscopic study of peroxidase localization in the onion root tip. Am. J. Bot. 62:280-291.

Goodman, R.N. 1967. Protection of apple stem tissue against *Erwinia amylovora* infection by avirulent strains and three other bacterial species. Phytopathology 57:22-24

Goodman, R.N. 1968. The hypersensitive reaction in tobacco: a reflection of changes in host cell permeability. Phytopathology 58:872-873.

Goodman, R.N. 1972. Electrolyte leakage and membrane damage in relation to bacterial population, pH and ammonia production in tobacco leaf tissue inoculated with *Pseudomonas pisi*. Phytopathology 62:1327-1331.

Goodman, R.N. 1976. Physiological and cytological aspects of the bacterial infection process. Pages 172-196 in: Encyclopedia of Plant Physiology, New Series, Vol. 4, Physiological Plant Pathology. R. Heitefuss and P.H. Williams, eds. Springer-Verlag, Berlin, Heidelberg, New York.

Goodman, R.N. and Burkowicz, A. 1970. Ultrastructural changes in apple leaves inoculated with a virulent and an avirulent strain of *Erwinia amylovora*. Phytopathol. Z. 68:258-268.

Goodman, R.N., Huang, P. Y., and White, J. A. 1976. Ultrastructural evidence for immobilization of an incompatible bacterium, *Pseudomonas pisi*, in tobacco leaf tissue. Phytopathol. 66:754-764.

Goodman, R.N., Kiraly, Z., and Wood K.R. 1986. The Biochemistry and Physiology of Plant Disease. Univ of Missouri Press.

Goodman, R.N. and Plurad, S.B. 1971. Ultrastructural changes in tobacco undergoing the hypersensitive reaction caused by plant pathogenic bacteria. Physiol. Plant Pathol. 1:11-15.

Gottlieb, D., and Shaw, P.D. 1970. Mechanism of action of antifungal antibiotics. Annu. Rev. Phytopathol. 8:371-402.

Gough, C.L., Genin, S., Zischek, C., and Boucher, C.A. 1992. *hrp* genes of *Pseudomonas solanacearum* are homologous to pathogenicity determinants of animal pathogenic bacteria and are conserved among plant pathogenic bacteria. Mol. Plant-Microbe Interact. 5:384-389.

Green, N.E., Hadwiger, L.A., and Graham, S.O. 1975. Phenylalanine ammonia-lyase, tyrosine ammonia-lyase, and lignin in wheat inoculated with *Erysiphe graminis* f. sp. *tritici*. Phytopathology 65:1071-1074.

Grimm, C., and Panopoulos, N.J. 1989. The predicted protein product of a pathogenicity locus from *Pseudomonas syringae* pv. *phaseolicola* is homologous to a highly conserved domain of several procaryotic regulatory proteins. J. Bacteriol. 171:5031-5038.

Gross, M. and Rudolph, K. 1987. Demonstration of levan and alginate in bean plants (*Phaseolus vulgaris*) infected by *Pseudomonas syringae* pv. *phaseolicola*. J. Phytopathol. (Berl.) 120:9-19.

Gross, S.S., Janse, C., and Elstner, E.F. 1977. Involvement of malate, monophenols and superoxide radical in hydrogen peroxide formation by the isolated cell walls from horseradish (*Amoracia lapathifolia* Gilib.) Planta 136:271-276.

Guern, J., and Ullrich-Eberius, C.I. 1988. The ferricyanide-driven redox system at the plasmalemma of plant cells: origin of the proton production and reappraisal of the stoichiometry e^-/H^+. Pages 253-262 in: Plasma Membrane Oxidoreductases in Control of Animal and Plant Growth. F.L. Crane, J.D. Morre, and H. Low, eds. Plenum Press, New York.

Gulyas, A., Barna, B., Klement, Z., and Farkas, G.L. 1979. Effect of plasmolytica on the hypersensitive reaction induced by bacteria in tobacco: a comparison with the virus-induced hypersensitive reaction. Phytopathology: 121-124.

Gulyas, A., and Farkas, G.L. 1978. Is cell-to-cell contact necessary for the expression of the N-Gene in *Nicotiana tabacum* cv. Xanthi nc. plants infected by TMV? Phytopathol. Z. 91:182-187.

Guo, A., and Leach, J.E. 1989. Examination of rice hydathode water pores exposed to *Xanthomonas*. Phytopathology. 79: 433-436.

Haberlach, G.T., Budde, A.D., Sequeira, L., Helgeson, J.P. 1978. Modification of disease resistance of tobacco callus tissues by cytokinins. Plant Physiol. 62:522-525.

Hadwiger. L.A., and Loschke, D.C. 1981. Molecular communication in host-parasite interactions: hexosamine polymers (chitosan) as regulator compounds in race-specific and other interactions. Phytopathology 71:756-762.

Halliwell, B. 1982. The Toxic Effects of Oxygen on Plant Tissuess. CRC Press, Boca Raton, FL. Pages 89-123.

Halliwell, B., and Gutteridge, J.M.C. 1984. Oxygen toxicity, oxygen radicals, transition metals and disease. Biochem. J. 219:1-14.

Halliwell, B., and Gutteridge, J.M.C. 1986. Iron and free radical reactions: two aspects of antioxidant protection. Trends Biochem. Sci. 11:372-375.

Hampson, S.E., Loomis, R.S. and Rains, D.W. 1978. Characteristics of sugar uptake in hypocotyls of cotton. Plant Physiology 62:846-850.

Hancock, J.G. 1977. Soluble metabolites in intercellular regions of squash hypocotyl tissues: Implications for exudation. Plant Soil 47:103-1

Hancock, J.G. and Huisman, O.C. 1981. Nutrient movement in host-pathogen systems. Annu. Rev. Phytopathol. 19:309-331.

Harder, D.E., Samborski, D.J., Rohringer, R., Rimmer, S., Kim, W., and Chong, J. 1979. Electron microscopy of susceptible and resistant near-isogenic (sr6/Sr6) lines of wheat are infected by *Puccinia graminis tritici* III Ultrastructure of incompatible reactions. Can. J. Bot. 57:2626-2634.

Hargreaves, J.A., and Barley, J.A. 1978. Phytoalexin production by metabolites released by damaged bean cells. Physiol. Plant Pathol. 13:89-100.

Hargreaves, J.A., and Selby, C. 1978. Phytoalexin formation in cell suspensions of *Phaseolus vulgaris* in response to an extract of bean hypocotyls. Phytochemistry 17:1099-1102.

Hasson, E.P., and Laties, G.G. 1976. Separation and characterization of potato lipid acylhydrolases. Plant Physiol. 57:142-147.

Hawker, J.S. 1965. The sugar content of cell walls and intercellular spaces in sugar-cane stems and its relation to sugar transport. Aust. J. Biol. Sci. 18:959-969.

Hayward, A.C. 1974. Latent infections by bacteria. Annu. Rev. Phytopathol. 12:87-97.

He, S.Y., Huang, H-C., and Collmer, A. 1993. *Pseudomonas syringae* pv. *syringae* $Harpin_{Pss}$: a protein that is secreted via the hrp pathway and elicits the hypersensitive response in plants. Cell 73: 1255-1266.

Heath, M.C. 1971. Haustorial sheath formation in cowpea leaves immune to rust infection. Phytopathology 61:383-388.

Heath, M. 1972. Ultrastructure of host and non-host reactions to cowpea rust. Phytopathology 62:27-38.

Heath, M.C. 1974. Light and electron microscope studies of the interaction of host and non-host plants with cowpea rust- *Uromyces phaseoli* var. *vignae*. Physiol. Plant Pathol. 4:403-414.

Heath, M.C. 1976. Hypersensitivity, the cause or the consequence of rust resistance? Phytopathology 66:935-936.

Heath, M.C. 1980. Reaction of non-suscepts to fungal pathogens. Annu. Rev. Phytopathol. 18:211-236.

Heath, M.C. 1982. Host defense mechanisms against infection by rust fungi, morphological, physiological and biochemical resistance to phytopathogens. Pages 223-245 in: The Rust Fungi. K.J. Scott and A.K. Chakravorty, eds. Academic Press, London, New York.

Heath, M.C. 1982. Fungal growth, haustorial disorganization, and host necrosis in two cultivars of cowpea inoculated with an incompatible race of the cowpea rust fungus. Physiol. Plant Pathol. 21: 347-359.

Heath, M.C., 1984. Relationship between heat-induced fungal death and plant necrosis in compatible and incompatible interactions involving the bean and cowpea rust fungi. Phytopathology 74:1370-1376.

Heath, M.C., 1988. Effect of fungal death or inhibition induced by oxycarboxin or polyoxin D on the interaction between resistant or susceptible bean cultivars and the bean rust fungus. Phytopathology 78:1454-1462.

Heath, M.C. 1989. A comparison of fungal growth and plant responses in cowpea and bean cultivars inoculated with urediospores or basidiospores of the cowpea rust fungus. Physiol. Mol. Plant Pathol. 34:415-426.

Heath, M.C., and Heath, I.B. 1971. Ultrastructure of an immune and a susceptible reaction of cowpea leaves to rust infection. Physiol. Plant Pathol. 1:277-287.

Hedrich, R., and Schroeder, J.I. 1989. The physiology of ion channels and electrogenic pumps in higher plants. Annu. Rev. Plant Physiol. 40:539-569.

Helgeson, J.P., Haberlach, G.T., Budde, A.D., Sequeira, L. 1976. Modification of disease resistance by kinetin. (Abstr.) Plant Physiol. 57:S-88.

Henfling, J.W.D.M., Bostock, R., and Kuc, J. 1980. Effect of abscisic acid on rishitin and lubimin accumulation and resistance of *Phytophthora infestans* and *Cladosporium cucumerinum* in potato tuber tissue slices. Phytopathology 70:1074-1078.

Hepler, P.K., and Wayne, R.O. 1985. Calcium and plant development. Annu. Rev. Plant Physiol. 36:397-439.

Herbers, K., Conrads-Strauch, J., and Bonas, U. 1992. Race-specificity of plant resistance to bacterial spot disease determined by repetitive motifs in a bacterial avirulence protein. Nature 356:172-174.

Higinbotham, N., 1973. Electropotentials of plant cells. Annu. Rev. Plant Physiol. 24:25-46.

Hignett, R.C. 1988. Effect of growth conditions on the surface structures and extracellular products of virulent and avirulent forms of *Erwinia amylovora*. Physiol. Mol. Plant Pathol. 32:387-394.

Hildebrand, D.C. 1973. Tolerance of homoserine by *Pseudomonas pisi* and implications of homoserine in plant resistance. Phytopathology 63:301-302.

Hildebrand, D.C., Alosi, M.C., and Schroth, M.N. 1980. Physical entrapment of pseudomonads in bean leaves by films formed at air-water interfaces. Phytopathology 70:98-109.

Hildebrand, D.C., and Riddle, B. 1971. Influence of environmental conditions on reaction induced by infiltration of bacteria into plant leaves. Hilgardia 41:33-43.

Hille, B. 1984. Ionic Channels of Excitable Membranes. Sinauer Associates, Inc., Sunderland, Massachusetts.

Hirano, S.S. and Upper, C.D. 1990. Population biology and epidemiology of *Pseudomonas syringae*. Annu. Rev. Phytopathol. 28:155-177.

Hohl, H.R., and Stossel, P. 1975. Host-parasite interfaces in a resistant and susceptible cultivar of *Solanum tuberosum* inoculated with *Phytophthora infestans*: tuber tissue. Can. J. Bot. 54:900-912.

Hohl, H.R., and Suter, E. 1976. Host-parasite interfaces in a resistant and a susceptible cultivar of *Solanum tuberosum* inoculated with *Phytophthora infestans*: leaf tissue. Can. J. Bot. 54:1956-1970.

Holliday, M.J., Keen, N.T., and Long, M. 1981. Cell death patterns and accumulation of fluorescent material in the hypersensitive response of soybean leaves to *Pseudomonas syringae* pv. *glycinea*. Physiol. Plant Pathol. 18:279-287.

Holmes, F.O. 1929. Local lesions in tobacco mosaic. Bot. Gaz. 87:39-55.

Holmes, F.O. 1938. Inheritance of resistance of the N gene to tobacco-mosaic disease in tobacco. Phytopathology 28:553-560.

Holmes, F.O. 1952. Use of primary lesions to determine relative susceptibility of individual plants. Phytopathology 42:113.

Hooley, R., and McCarthy, D. 1980. Extracts from virus infected hypersensitive tobacco leaves are detrimental to protoplast survival. Physiol. Plant Pathol. 16:25-38.

Horikawa, T., Tomiyama, K., and Doke, N. 1976. Accumulation and transformation of rishitin and lubimin in potato tuber tissue infected by an incompatible race of *Phytophthora infestans*. Phytopathology 66:1186-1191.

Horn, M.A., Heinstein, P.F., and Low, P.S. 1989. Receptor-mediated endocytosis in plant cells. Plant Cell 1:1003-1009.

Hsieh, S.P.Y., and Buddenhagen, I.W. 1974. Suppressing effects of *Erwinia herbicola* on infection by *Xanthomonas oryzae* and on symptom development in rice. Phytopathology 64: 1182-1185.

Huang, H.-C., Hutcheson, S.W., and Collmer, A. 1991. Characterization of the *hrp* cluster from *Pseudomonas syringae* pv. *syringae* 61 and Tn*phoA* tagging of genes encoding exported or membrane-spanning Hrp proteins. Mol. Plant-Microbe Interact. 4:469-476.

Huang, J.C., Schuurink, R., Denny, T.P., Atkinson, M.M., Baker, C.J., Yucel, I., Hutcheson, S.W., and Collmer, A. 1988. Molecular cloning of a *Pseudomonas syringae* pv. *syringae* gene cluster that enable *Pseudomonas fluorescens* to elicit the hypersensitive response in tobacco plants. J. Bacteriol. 170:4748-4756.

Huang, J.S., and Goodman, R.N. 1972. Recognition of pathogenic and saprophytic bacteria by tobacco leaf cells, reflected as changes in respiratory rates. Acta Phytopathol. Acad. Sci. Hung. 20:7-15.

Huang, Y., Helgerson, J.P., and Sequeira, L. 1989. Isolation and purification of a factor from *Pseudomonas solanacearum* that induces a hypersensitive-like response in potato cells. Mol. Plant-Microbe Interact 2:132-138.

Huang, Y., Xu, P., and Sequeira, L. 1990. A second cluster of genes that specify pathogenicity and host response in *Pseudomonas solanacearum*. Mol. Plant-Microbe Interact 3:48-53.

Hunter, M.I.S., Hetherington, A.M., and Crawford, M.M. 1983. Lipid peroxidation - a factor in anoxia intolerance in iris species? Phytochemistry 22:1145-1147.

Husken, D., Steudle, E., and Zimmermann, U. 1978. Pressure probe technique for measuring water relations of cells in higher plants. Plant Pysiol. 61:158-163.

Hutcheson, S.W., Collmer, A., and Baker, C.J. 1989. Elicitation of the hypersensitive response by *Pseudomonas syringae*. Physiol. Plant. 76:155-163.

Hutchenson, S.W., Heu, S., Huang, H.-C., Li, T.-H., Lu, Y., and Xiao, Y. 1993. The *hrp* genes of *Pseudomonas syringae*: an evaluation of their role in determining non-host resistance. In: Biotechnology and Plant Protection, Bacterial Pathogenesis and Disease Resistance. D. Bills and S.D. Kung, eds. Butterworth Publishers (In press)

Huynh, T.V., Dahlbeck, D., Staskawicz, B.J. 1989. Bacterial blight of soybean: regulation of a pathogen gene determining host cultivar specificity. Science 245:1374-1477.

Ingram, D.S., Sargent, J.A., and Tommerup, I.C. 1976. Structural aspects of infection by biotrophic fungi. Pages 43-78 in: Biochemical Aspects of Plant-parasite Relationships. J. Friend and D. Threlfall, eds. Academic Press, London.

Ishiguri, Y., Tomiyama, K., Doke, N., Murai, A., Katsui, N., Yagihashi, F., and Masamune, T. 1978. Induction of rishitin-metabolizing activity in potato tuber tissue disks by wounding and identification of rishitin metabolites. Physiol. Biochem. 68:720-725.

Ishizaka, N., Matsunaga, N., Imaizumi, K., and Matsumune, T. 1969. Biological activities of rishitin, an antifungal compound isolated from diseased potato tubers, and its derivative. Plant Cell Physiol. 10:183-192.

Ishizaka, N., and Tomiyama, K. 1970. Effect of *Phytophthora infestans* infection on wound periderm formation in cut potato tuber tissue. Ann. Phytopathol. Soc. Jpn. 36:243-249.

Ishizaka, N., and Tomiyama, K. 1972. Effect of wounding or infection by *Phytophthora infestans* on the contents of terpenoids in potato tubers. Plant Cell Physiol. 13:1053-1063.

Israel, H.W., and Ross, A.F. 1967. The fine structure of local lesions induced by tobacco mosaic virus in tobacco. Virology 33:272-286.

Jakobek, J.L., and Lindgren, Peter B. 1993. Generalized induction of defense responses in bean is not correlated with the induction of the hypersensitive reaction. Plant Cell. 5:49-56.

Jakobek, J.L., Smith, J.A., and Lindgren, P.B. 1993. Suppression of bean defense responses by *Pseudomonas syringae*. Plant Cell. 5:57-63.

John, V.T., and Weintraub, M. 1967. Phenolase activity in *Nicotiana glutinosa* infected with tobacco mosaic virus. Phytopathology 57:154-158.

Jones, S.B., and Fett, W.F. 1985. Fate of *Xanthomonas campestris* infiltrated into soybean leaves: An ultrastructural study. Phytopathology 75:733-741.

Jordan, C.M., and DeVay, J.E. 1990. Lysosome disruption associated with hypersensitive reaction in the potato-*Phytophthora infestans* host-parasite interaction. Physiol. Mol. Plant Pathol. 36:221-236.

Kamiya, N. 1981. Physical and chemical basis of cytoplasmic streaming. Annu. Rev. Plant Physiol. 32:205-236.

Kao, C.C., and Sequeira, L. 1991. A gene cluster required for coordinated biosynthesis of lipopolysaccharide and extracellular polysaccharide also affects virulence of *Pseudomonas solanacearum*. J. Bacteriol. 173:7841-7847.

Kato, S., and Misawa, T. 1976. Lipid peroxidation during the appearance of hypersensitive reaction in cowpea leaves infected with cucumber mosaic virus. Ann. Phytopathol. Soc. Jpn. 42:472-480.

Kauss, H. 1990. Role of the plasma membrane in host-pathogen interactions. Pages 320-350 in: The Plant Plasma Membrane. Structure, Function and Molecular Biology, C. Larson and I.M. Moller, eds. Springer-Verlag, Heidelburg, Berlin.

Kavanaugh, T., Goulden, M., Santa Cruz, S., Chapman, S., Barker, I., and Baulcombe, D. 1992. Molecular analysis of a resistance-breaking strain of Potato Virus X. Virology 189:609-617.

Keen, N.T. 1975. Specific elicitors of plant phytoalexin production: Determinants of race specificity in pathogens? Science 187:74-75.

Keen, N. 1990. Gene-for-gene complementarity in plant-pathogen interactions. Annu. Rev. Genet. 24:447-463.

Keen, N.T., Ersek, R., Long, M., Bruegger, B., and Holliday, M. 1981. Inhibition of the hypersensitive reaction of soybean leaves to incompatible *Pseudomonas* sp by blasticidin S, streptomycin or elevated temperature. Physiol. Plant Pathol. 18:325-337.

Keen, N.T., and Kenney, B.W. 1974. Hydroxyphaseollin and related isoflavanoids in the hypersensitive resistance reactions of soybeans to *Pseudomonas glycinea*. Physiol. Plant Pathol. 4:173-185.

Keen, N.T., and Staskawicz, B. 1988. Host range determinants in plant pathogens and symbionts. Annu. Rev. Microbiol. 42:421-40.

Keen, N.T., Tamaki, S., Kobayashi, D., Gerhold, D., Stayton, M., Shen, H., Gold, S., Lorang, J., Thordal-Christensen, H., Dahlbeck, D., Staskawicz, B. 1990. Bacteria expressing avirulence gene D produce a specific elicitor of the soybean hypersensitive reaction. Mol. Plant-Microbe Interact. 3:112-121.

Kelman, A. 1954. The relationship of pathogenicity of *Pseudomonas solanacearum* to colony appearance on a tetrazolium medium. Phytopathology 44:693-695.

Kelman, A., and Cowling, E.B. 1965. Cellulase of *Pseudomonas solanacearum* in relation to pathogenesis. Phytopathology 55:148-155.

Kelman, A., and Hruscka, H. 1973. The role of motility and aerotaxis in the selective increase of avirulent bacteria in still broth cultures of *Pseudomonas solanacearum*. J. Gen. Microbiol. 76:177-188.

Kennedy, B.W., and Ercolani, G.L. 1978. Soybean primary leaves as a site for epiphytic multipication of *Pseudomonas glycinea*. Phytopathology 68:1196-1201.

Keppler, L.D., Atkinson, M.M., and Baker, C.J. 1988. Plasma membrane alteration during bacteria-induced hypersensitive reaction in tobacco suspension cells as monitored by intracellular accumulation of fluorescein. Physiol. Mol. Plant Pathol. 32:209-219.

Keppler, L.D., and Baker, C.J. 1989. O_2^- -initiated lipid peroxidation in a bacteria-induced hypersensitive reaction in tobacco cell suspensions. Phytopathlogy 79:555-562.

Keppler, L.D., Baker, J.C., and Atkinson, M.M. 1989. Active oxygen production during a bacteria-induced hypersensitive reaction in tobacco suspension cells. Phytopathology 79:974-978.

Keppler, L.D., and Novacky, A. 1986. Involvement of membrane lipid peroxidation in the development of a bacterially induced hypersensitive reaction. Phytopathology 76:104-108.

Keppler, L.D., and Novacky, A. 1987. The initiation of membrane lipid peroxidation during bacteria-induced hypersensitive reaction. Physiol. Mol. Plant Pathol. 30:233-245.

Keppler, L.D., and Novacky, A. 1989. Changes in cucumber cotyledon membrane lipid fatty acids during paraquat treatment and a bacteria-induced hypersensitive reaction. Phytopathology 79:705-708.

Kiraly, Z. 1980. Defense triggered by the invader: Hypersensitivity. Pages 201-224 in: Plant Disease: An Advanced Treatise. Vol 5. J.G. Horsfall and E.B. Cowling, eds., Academic Press Inc. New York.

Kitazawa, K., Inagaki, H., and Tomiyama. K. 1973. Cinephotomicrographic observations on the dynamic responses of protoplasm of a potato plant cell to infection by *Phytophthora infestans*. Phytopathol. Z. 76:80-86.

Kitazawa, K., and Tomiyama, K. 1969. Microscopic observations of infection of potato cells by compatible and incompatible races of *Phytophthora infestans*. Phytopathol. Z. 66:317-324.

Kitazawa, K., and Tomiyama, K. 1973. Role of underlying healthy tissue in the hypersensitive death of a potato plant cell infected by an incompatible race of *Phytophthora infestans.* Ann. Phytopathol. Soc. Jpn. 39:85-59.

Klement, Z. 1963. Method for the rapid detection of the pathogenicity of phytopathogenic pseudomonads. Nature 199:299-300.

Klement, Z. 1965. Method of obtaining fluid from the intercellular space of foliage and the fluid's merit as substrate for phytobacterial pathogens. Phytopathology 55:1033-1034.

Klement, Z. 1977. Cell contact recognition versus toxin action in induction of bacterial hypersensitive reaction. Acta Phytopathol. Acad. Sci. Hung. 12:257-261.

Klement, Z. 1982. Hypersensitivity. Pages 149-177 in: Phytopathogenic Prokaryotes. Vol 2. M.S. Mount and G.H. Lacy, eds. Academic Press, New York.

Klement, Z., Farkas, G. L., and Lovrekovich, L. 1964. Hypersensitive reaction induced by phytopathogenic bacteria in the tobacco leaf. Phytopathology 54:474-477.

Klement, Z., and Goodman, R.N. 1967a. The role of the living bacterial cell and induction time in the hypersensitive reaction of the tobacco plant. Phytopathology 57:322-323.

Klement, Z., and Goodman, R.N. 1967b. The hypersensitive reaction to infection by bacterial plant pathogens. Annu. Rev. Phytopathol. 5:17-44.

Klement, Z., and Lovrekovich L. 1961. Studies on host-parasite relations in bean pods infected with bacteria. Phytopathol. Z. 45:81-88.

Klement, Z., Rudolph, K., and Gross, M. 1987. Necrosis instead of water-soaking due to light deficiency in leaves after inoculation with pseudomonads or xanthomonads. Pages 530-536 in: Plant Pathogenic Bacteria Proc. VI Int. Conf. Plant Path. Bact. 1985, Maryland, USA. E.L. Civerolo, A. Collmer, R.E. Davis, and A.G. Gillaspie, eds., Martinus Nijhoff Publishers, Dordrecht, Boston, Lancaster.

Knoop, V., Staskawicz, B., and Bonas, U. 1991. Expression of the avirulence gene *avrBs3* from *Xanthomonas campestris* pv. *vesicatoria* is not under the control of *hrp* genes and is independent of plant factors. J. Bacteriol. 173:7142-7150.

Knorr, D.A., and Dawson, W.O. 1987. A point mutation in the tobacco mosaic virus capsid protein gene induces hypersensitivity in *Nicotiana sylvestris.* Proc. Natl. Acad. Sci. U.S.A. 85:170-174.

Kobayashi, D.Y., and Keen, N.T. 1986. Analysis of an avirulence gene from *Pseudomonas syringae* pv. *tomato* which elicits hypersensitive response on soybean. (Abstr.) Phytopathology 76:1099.

Kobayashi, D.Y., Tamaki, S.J., and Keen, N.T. 1990. Molecular characterization of avirulence gene D from *Pseudomonas syringae* pv. *tomato*. Mol. Plant-Microbe Interact. 3:94-102.

Koch, E., Meier, B.M., Eiben, H.G., Slusarenko, A. 1992. A lipoxgenase from leaves of tomato (*Lycopersicon esculentum* Mill.) is induced by response to plant pathogenic pseudomonads. Plant Physiol. 99:571-576.

Konate, G., and Fritig, B. 1984. An efficient microinoculation procedure to study plant virus multiplication at predetermined individual infection sites on the leaves. Phytopathol. Z. 109:131-138.

Kontaxis, D.G. 1961. Movement of tobacco mosaic virus through epidermis of *Nicotiana glutinosa* leaves. Nature 192:581-582.

Konze, J.R., Elstner, E.F. 1978. Ethane and ethylene formation by mitochondria as indication of aerobic lipid degradation in response to wounding of plant tissue. Biochim. Biophys. Acta 528: 213-221.

Kopp, M., Geoffroy, P. and Fritig, B. 1979. Phenylalanine ammonia-lyase levels in protoplasts isolated from hypersensitive tobacco pre-infected with tobacco mosaic virus. Planta 146:451-457.

Kubowitz, B.D., Vanderhoff, L.N., and Hanson, J.B. 1982. ATP-dependent calcium transport in plasmalemma preparations from soybean hypocotyls. Plant Physiol. 69:187-191.

Kunoh, H., Aist, J.R., and Hashimoto, A. 1985. The occurrence of cytoplasmic aggregates induced by *Erysiphe pisi* in barley coleoptile cells before the host cell walls are penetrated. Physiol. Plant Pathol. 26:199-207.

Kursanov, A.L., and Brovchenko, M.I. 1970. Sugars in the free space of leaf plates: their origin and possible involvement in transport. Can. J. Bot. 48:1243-1250.

Ladyzhenskaya, E.P., Adam, A., Korableva N.P., and Ersek, T. 1991. Plasmalemma-bound ATPase of potato tubers and its inhibition by *Phytophthora infestans*-derived elicitors of the hypersensitive reaction. Plant Sci. (Shannon) 73:137-142.

Lauchli, A. 1976. Apoplasmic transport in tissues. Pages 3-34 in: Encyclopedia of Plant Physiology, New Series Vol. 2, part B. U. Luttge and M.G. Pitman, eds. Springer-Verlag, Berlin, Heidelberg, New York.

Lawton, M.A., Dixon, R.A., Hahlbrock, K., and Lamb, C.J. 1983. Rapid induction of the synthesis of phenylalanine ammonia-lyase and chalcone synthase in elicitor-treated plant cells. Eur. J. Biochem. 129:593-601.

Lawton, M.A., and Lamb, C.J. 1987. Transcriptional activation of plant defense genes by fungal elicitor, wounding, and infection. Mol. Cell. Biol. 7:335-341.

Lea, P.J., Blackwell, R.D., and Joy, K.W. 1992. Ammonia assimilation in higher plants. Pages 153-186 in: Nitrogen Metabolism of Plants. K. Mengel and D.J. Pilbeam, eds. Clarendon Press, Oxford.

Leath, K.T., and Rowell, J.B. 1966. Histological study of the resistance of *zea mays* to *Puccinia graminis*. Phytopathology 56:1305-1309.

Leatham, G.F., King, V., and Stahmann, M.A. 1980. In vitro protein polymerization by quinones or free radicals generated by plant or fungal oxidative enzymes. Phytopathology 70:1134-1140.

Leben, C. 1965. Epiphytic microorganisms in relation to plant disease. Annu. Rev. Phytopathol. 3:209-230.

Leben, C. 1981. How plant-pathogenic bacteria survive. Plant Dis. 65:633-637.

Leben, C., Schroth, M.N., and Hildebrand, D.C. 1970. Colonization and movement of *Pseudomonas syringae* on healthy bean seedlings. Phytopathology 60:677-680.

Leben, C., and Whitmoyer, R.E. 1979. Adherence of bacteria to leaves. Can. J. Microbiol. 25:896-901.

Legrand, M. 1983. Phenylpropanoid metabolism and its regulation in disease. Pages 367-384 in: Biochemical Plant Pathology. J.A. Callow, ed. John Wiley & Sons Ltd., New York.

Legrand, M., Fritig, B., and Hirth, L. 1976. Enzymes of the phenylpropanoid pathway and the necrotic reaction of hypersensitive tobacco to tobacco mosaic virus. Phytochemistry (Oxf.) 15:1353-1359.

Leshem, Y.Y. 1981. Oxy free radicals and plant senescence. Whats New Plant Physiol. 12:1-4.

Leshem, Y.Y. 1984. Interaction of cytokinins with lipid-associated oxy free radicals during senescence: a prospective mode of cytokinin action. Can. J. Bot. 62:2943-2949.

Leshem, Y.Y. 1987. Membrane phospholipid catabolism and Ca^{2+} activity in control of senescence. Physiol. Plant. 69:551-559.

Leshem, Y.Y. 1988. Plant senescence processes and free radicals. Free Radical Biol. & Med. 5:39-49.

Leshem, Y.Y., and Bar-nes, G. 1982. Lipoxygenase as affected by free radical metabolism: senescence retardation by xanthine oxidase inhibitor allopurinol. Pages 275-278 in: Biochemistry and Metabolism of Plant Lipids. J.F. Wintermans, P.J.C. Kurper, eds. Elsevier, Amsterdam.

Li, T.H., Benson, S.A., and Hutcheson, S.W. 1992. Phenotypic expression of the *Pseudomonas syringae* pv. *syringae* 61 *hrp/hrm* gene cluster in *Escherichia coli* MC4100 requires a functional porin. J. Bacteriol. 174: 1742-1749.

Lindbeck, D.A.C., Dawson, W.O., and Thomson, W.W. 1991. Coat protein-related polypeptides from in vitro tobacco mosaic virus mutants do not accumulate in chloroplasts of directly inoculated leaves. Mol. Plant-Microb. Interact. 4:89-94.

Lindgren, P.B., Panopoulos, N.J., Staskawicz, B.J., and Dahlbeck, D. 1988. Genes required for pathogenicity and hypersensitivity are conserved and interchangeable among pathovars of *Pseudomonas syringae.* Mol. Gen. Genet. 211:499-506.

Lindgren, P.B., Peet, R.C., and Panopoulos, N.J. 1985. Cloning and analysis of genes associated with pathogenicity and hypersensitivity from *Pseudomonas syringae* pv *phaseolicola*. Phytopathology 75: 1355.

Lindgren, P.B., Peet, R.C., and Panopoulos, N.J. 1986. Gene cluster of *Pseudomonas syringae* pv. "*phaseolicola*" controls pathogenicity on bean plants and hypersensitivity on nonhost plants. J. Bacteriol. 168:512-522.

Lisker, N., and Kuc, J. 1977. Elicitors of terpenoid accumulation in potato slices. Phytopathology 67:1356-1359.

Littlefield, L.J. 1973. Histological evidence for diverse mechanisms of resistance to flax rust, *Melampsora lini* (Ehrenb.) Lev. Physiol. Plant Pathol. 3:241-247.

Loebenstein, G. 1972. Localization and induced resistance in virus infected plants. Annu. Rev. Phytopathol. 10:171-206.

Loebenstein, G., Gera, A., and Gianinazzi, S. 1990. Constitutive production of an inhibitor of virus replication in the interspecific hybrid of *Nicotiana glutinosa* X *Nicotiana debneyi*. Physiol. Mol. Plant Pathol. 37:145-151.

Long, S.R. 1989. Rhizobium-legume nodulation: life together in the underground. Cell 56:203-214.

Lovrekovich, L., and Farkas, G.L. 1965. Induced protection against wildfire disease in tobacco leaves treated with heat-killed bacteria. Nature 205:823-824.

Lovrekovich, L., Lovrekovich, H., and Goodman, R.N. 1970. Ammonia as a necrotoxin in the hypersensitive reaction caused by bacteria in tobacco leaves. Can. J. Bot. 48:167-171.

Lovrekovich, L., and Lovrekovich, H. 1970. Tissue necrosis in tobacco caused by a saprophytic bacterium. Phytopathology 60:1279-1280.

Low, P.S., and Heinstein, P.F. 1986. Elicitor stimulation of the defense response in cultured plant cells monitored by fluorescent dyes. Arch. Biochem. Biophys. 249:472-279.

Lozano, J.C., and Sequeira, L. 1970a. Differentiation of races of *Pseudomonas solanacearum* by a leaf infiltration technique. Phytopathology 60:833-838.

Lozano, J.C., and Sequeira, L. 1970b. Prevention of hypersensitive reaction in tobacco leaves by heat-killed bacterial cells. Phytopathology 60:875-879.

Lyon, F., and Wood, R.K.S. 1976. The hypersensitive reaction and other responses of bean leaves to bacteria. Ann. Bot. (Lond) 40:479-491.

Maclean, D.J., Sargent, J.A., Tommerup, I.C., Ingram, D.S. 1974. Hypersensitivity as the primary event in resistance to fungal parasites. Nature 249:186-187.

Mader, M., Nessel, A., and Schloss, P. 1986. Cell compartmentation and specific roles of isoenzymes. Pages 247-259 in: Molecular and Physiological Aspects of Plant Peroxidases. H. Greppin, C. Penel, and T. Gaspar, eds. Univ. of Geneva, Switzerland.

Malik, A.N., Vivian, A., and Taylor, J.D. 1987. Isolation and partial characterization of three classes of mutants in *Pseudomonas syringae* pathovar *pisi* with altered behavior towards their host *Pisum sativum*. J. Gen. Microbiol. 133:2393-2399.

Maniara, G., Laine, R., and Kuc. J. 1984. Oligosaccharides from *Phytophthora infestans* enhance the elicitation of sesquiterpenoid stress metabolites by arachidonic acid in potato. Physiol. Plant Pathol. 24:177-186.

Mansfield, J.W., and Sexton, R. 1974. Changes in the localization of B-glycerophosphatase activity during the infection of *Phaseolus vulgaris* by *Colletotrichum lindemuthianum*. Ann. Bot. 38:711-17.

Marme, D. 1989. The role of calcium and calmodulin in signal transduction. Pages 57-80 in: Second Messengers in Plant Growth and Development. W.F. Boss and D.J. Morre, eds. Alan R. Liss, Inc., New York.

Marryat, D.C.E. 1907. Notes on the infection and histology of two wheats immune to the attacks of *Puccinia glumarum*, yellow rust. J. Agr. Sci. 2:129-138.

Matsumoto, N., Tomiyama, K., and Doke, N. 1976. Alteration of membrane permeability of potato-tuber tissue infected by incompatible and compatible races of *Phytophthora infestans* during the initial phase of infection. Ann. Phytopathol. Jpn. 42:279-286.

Matthysse, A.G. 1987. A method for the bacterial elicitation of a hypersensitive-like response in plant cell cultures. J. Microbiol. Methods 7:183-191.

Mauch, F., and Staehelin, A. 1989. Functional implications of the subcellular localization of ethylene-induced chitinase and β-1,3-glucanase in bean leaves. Plant Cell 1:447-457.

Mazzucchi, U., and Pupillo, P. 1976. Prevention of confluent hypersensitive necrosis in tobacco leaves by a bacterial protein-lipopolysaccharide complex. Physiol. Plant Pathol. 9:101-112.

McRobbie, E.A.C. 1987. Ionic relations of guard cells. Pages 125-162 in: Stomatal Function. E. Zeiger, G.D. Farquhar, and I.R. Cowan, eds. Stanford University Press, Stanford, CA.

Meadows, M.E., and Stall, R.E. 1981. Different induction periods for hypersensitivity in pepper to *Xanthomonas vesicatoria* determined with antimicrobial agents. Phytopathology 71:1024-1027.

Mehlhorn, H., and Kunert, K.J. 1986. Ascorbic acid, phenolic compounds, and plant peroxidases: a natural defence system against peroxidative stress in higher plants? Pages 437-440 in: Molecular and Physiological Aspects of Plant Peroxidases. H. Greppin, C. Penel, and T. Gaspar, eds. Univ. of Geneva, Switzerland.

Mendgen, K. 1975. Ultrastructural demonstration of different peroxidase activities during the bean rust infection process. Physiol. Plant Pathol. 6:275-282.

Meshi, T., Motoyoshi, F., Adachi, A., Watanabe, H., Takamatsu, N., and Okada, Y. 1988. Two concomitant base substitutions in the putative replicase genes of tobacco mosaic virus confer the ability to overcome the effects of the tomato resistance gene, Tm-1. Eur. Mol. Biol. Organ. J. 7:1575-1581.

Meyer, S.L.F., and Heath, M.C. 1988a. A comparison of the death induced by fungal invasion or toxic chemicals in cowpea epidermal cells. I. Cell death induced by heavy metal salts. Can. J. Bot. 66:613-623.

Meyer, S.L.F., and Heath, M.C. 1988b. A comparison of the death induced by fungal invasion or toxic chemicals in cowpea epidermal cells. II. Responses induced by *Erysiphe cichoracearum*. Can. J. Bot. 66:624-634.

Miflin, B.J., and Lea, P.J. 1982. Ammonia assimilation and amino acid metabolism. Pages 5-64 in: The Encyclopedia of Plant Physiology, vol. 14A. D. Boultes and B. Parthier, eds. Springer Verlag, Berlin.

Mindrinos, M.N., Rahme, L.G., Fredrick, R.D., Hatziloukas, E., Grimm, C., and Panopoulos, N.J. 1990. Stucture, function, regulation, and evolution of genes involved in pathogenicity, the hypersensitive response and phaseolotoxin immunity in the bean halo blight pathogen. Pages 74-81 in: *Pseudomonas*: Biotransformations, Pathogenesis and Evolving Biotechnology. S. Silver, A.M. Chakrabarty, B. Inglewski, and S. Kaplan, eds. American Society for Microbiology, Washington, D.C.

Mishra, O.P., Delivoria-Papadopoulos, M., Cahillane, G., and Wagerle, L.C. 1989. Lipid peroxidation as the mechanism of modification of the affinity of the Na^+, K^+-ATPase active sites for ATP, K^+, Na^+ and strophanthidin *in vitro*. Neurochem. Res. 14:845-851.

Mo, Y.Y., and Gross, D.C. 1991. Plant signal molecules activate the *syrB* gene, which is required for syringomycin production by *Pseudomonas syringae* pv. *syringae*. J. Bacteriol. 173: 5784-5792.

Mobley, R.D., Schroth, M.N., and Hildebrand, D.C. 1972. 35 sulfur diffusion from bean leaves in relation to growth of *Pseudomonas phaseolicola*. Phytopathology 62:300-301.

Modderman, P.W., Schot, C.P., Klis, F.M., and Wieringer-Brants, D.H. 1985. Aquired resistance in hypersensitive tobacco against tobacco mosaic virus, induced by cell wall components. Phytopathol. Z. 113:165-170.

Moerschbacher, B.M., Noll, U., Gorrichon, L., and Reisener, H.J. 1990. Specific inhibition of lignification breaks hypersensitive resistance of wheat to stem rust. Plant Physiol. 93:465-470.

Mohan, R., and Kolattukudy, P.E. 1990. Differential activation of expression of a suberization-associated anionic peroxidase gene in near-isogenic resistant and susceptible tomato lines by elicitors of *Verticillium albo-atrum*. Plant Physiol. 921:276-280.

Moreau, R.A., and Osman, S.F. 1989. The properties of reducing agents released by treatment of *Solanum tuberosum* with elicitors from *Phytophthora infestans*. Physiol. Mol. Plant Pathol. 35:1-10.

Morgham, A.T., Richardson, P.E., Essenberg, M., and Cover, E.C. 1988. Effects of continuous dark upon ultrastructure, bacterial populations and accumulation of phytoalexins during interactions between *Xanthomonas campestris* pv. *malvacearum* and bacterial blight susceptible and resistant cotton. Physiol. Mol. Plant Pathol. 32:141-162.

Morris, E.R., Rees, D.A., Young, G., Walkinshaw, M.D., and Darke, A. 1977. Order-disorder transition for a bacterial polysaccharide in solution. A role for polysaccharide conformation in recognition between *Xanthomonas* pathogen and its plant host. J. Mol. Biol. 110:1-16.

Mukherjee, D., Lambert, J.W., Cooper, R.L., and Kennedy, B.W. 1966. Inheritance of resistance to bacterial blight in soybean. Crop Sci. 6:324-326.

Muller, K.O. 1959. Hypersensitivity. Pages469-510 in: Plant Pathology. Vol. 1. J.S. Horsfall and A.E. Dimond, eds. Academic Press, New York.

Muller, K.O., and Behr, L. 1949. Mechanism of *Phytophthora* resistance of potatoes. Nature 163:498-499.

Murai, A. 1987. Phytoalexin chemistry and action. Pages 81-88 in: Pesticide Science and Biotechnology. R. Greenhalgh and T.R. Roberts, eds. Blackwell Sci. Publ. Oxford.

Nakagaki, Y., Hirai, T., and Stahmann, M.A. 1970. Ethylene production by detached leaves infected with tobacco mosaic virus. Virology 40:1-9.

Nakajima, T., Tomiyama, K., and Kinukawa, M. 1975. Distribution of rishitin and lubimin in potato-tuber tissue infected by an incompatible race of *Phytophthora infestans* and the site where rishitin is synthesized. Ann. Phytopathol. Soc. Jpn. 41:49-55.

Napoli, C., and Staskawicz, B.J. 1987. Molecular characterization of an avirulence gene from Race 6 of *Pseudomonas syringae* pv. *glycinea*. J. Bacteriol. 169:572-578.

Neukom, H., and Markwalder, H.U. 1978. Oxidative gelation of wheat and flour pentosans: A new way of cross-linking polymers. Cereal Foods World 23:374-376.

Newcombe, G., and Robb, J. 1989. The chronological development of a lipid-to-suberin response at *Verticillium* trapping sites in alfalfa. Physiol. Mol. Plant Pathol. 34:55-73.

Niepold, F., Anderson, D., and Mills, D. 1985. Cloning determinants of pathogenesis from *Pseudomonas syringae* pathovar *syringae*. Proc. Natl. Acad. Sci. USA 82:406-410.

Nishiguchi, M., Langridge, W.H.R., Szalay, A.A., and Zaitlin, M. 1986. Electroporation mediated infection of tobacco leaf protoplasts with tobacco mosaic virus RNA and cucumber mosaic virus RNA. Plant Cell Rep. 5:57-60.

Nishimura, N., Tomiyama, K., and Doke, N. 1978. Effect of infection by *Phytophthora infestans* and treatment with its zoosporial components on uptake of ^{3}H-leucine and protein synthesis in potato tuber tissue. Ann. Phytopathol. Soc. Jpn. 44:599-605.

Nishimura, N., and Tomiyama, K. 1978. Effect of infection by *Phytophthora infestans* on uptake of ^{3}H-leucine, ^{32}P and ^{86}Rb by potato-tuber tissue. Ann. Phytopathol. Soc. Jpn. 44:159-166.

Nonami, H., and Boyer, J.S. 1990. Wall extensibility and cell hydraulic conductivity decrease in enlarging stem tissues at low water potentials. Plant Physiol. 93:1610-1619.

Novacky, A. 1972. Suppression of the bacterially induced hypersensitive reaction by cytokinins. Physiol. Plant Pathol. 2:101-104.

Novacky, A. 1980. Disease-related alteration in membrane function. Pages 369-380 in: Proceedings International Workshop held in Toronto, Canada, July 22-27, 1979. J. Dainty and R. Spanswick, eds. Elsevier/North Holland Biomedical Press.

Novacky, A., Acedo, G., and Goodman, R.N. 1973. Prevention of bacterially induced hypersensitive reaction by living bacteria. Physiol. Plant Pathol. 3:133-136.

Novacky, A., and Hanchey, P. 1976. Effect of internal injury on bacterial hypersensitivity. Acta Phytopathol. Acad. Sci. Hung. 11:217-222.

Novacky, A., Karr, A., and Van Sambeek, J.W. 1976. Using electrophysiology to study plant disease development. BioScience 26:499-504.

Novacky, A., Pike, S.M., and Popham, P.L. 1993. Varying membrane properties of cells in tissues inoculated with HR-inducing bacteria. In D. Bills and S. D. Kung, eds. Biotechnology and Plant Protection. Bacterial Pathogenesis and Disease Resistance. Butterworth Publishers (In press)

Novacky, A., and Ullrich-Eberius, C.T. 1982. Relationship between membrane potential and ATP level in *Xanthomonas campestris* pv. *malvacearum* infected cotton cotyledons. Physiol. Plant Pathol. 21:237-249.

Nozue, M., Tomiyama, K., and Doke, N. 1978. Effect of adenosine 5'-triphosphate on hypersensitive death of potato tuber cells infected by *Phytophthora infestans*. Phytopathology 68:873-876.

Nozue, M., Tomiyama, K., and Doke, N. 1979. Evidence for adherence of host plasmalemma to infecting hyphae of both compatible and incompatible races of *Phytophthora infestans*. Physiol. Plant Pathol. 15:111-115.

Nozue, M., Tomiyama, K., and Doke, N. 1980a. Effect of some enzyme inhibitors on ATP level and hypersensitive reactivity of potato tuber disks to incompatible race of *Phytophthora infestans*. Ann. Phytopath. Soc. Jpn. 46:34-39.

Nozue, M., Tomiyama. K., and Doke, N. 1980b. Induction of hypersensitive response in potato tuber tissue to compatible race of *Phytophthora infestans* by treatment with amino sugars: conversion from susceptible to resistant tissue. Ann. Phytopathol. Soc. Jpn. 46:250-252.

Nozue, M., Tomiyama, K., and Doke, N. 1980c. Effect of N, N'-diacetyl-D-chitobiose the potato-lectin hapten and other sugars on hypersensitive reaction of potato tuber cells infected by incompatible and compatible races of *Phytophthora infestans*. Physiol. Plant Pathol. 17:221-227.

Nozue, M., Tomiyama, K., and Doke, N. 1981. Effect of some enzyme inhibitors on adherence of host plasmalemma to infecting hyphae of *Phytophthora infestans*. Ann. Phytopathol. Soc. Jpn. 47:189-193.

O'Brien, F., and Wood, R.K.S. 1973. Role of ammonia in infection of *Phaseolus vulgaris* by *Pseudomonas* spp. Physiol. Plant Pathol. 3:315-326.

O'Connell, K.P., and Handelsman, J. 1989. *chvA* locus may be involved in export of neutral cyclic ß-1,2-linked D-glucan from*Agrobacterium tumefaciens*. Mol. Plant-Microbe Interact. 2:11-16.

Ohashi, Y., and Shimomura, T. 1976. Leakage of cell constituents associated with local lesion formation on *Nicotiana glutinosa* leaf infected with tobacco mosaic virus. Ann. Phytopathol. Soc. Jpn. 42:436-441.

Ohashi, Y., and Shimomura, T. 1982. Modification of cell membranes of leaves systemically infected with tobacco mosaic virus. Physiol. Plant Pathol. 19:125-128.

Omer, M.E.H., and Wood, R.K.S. 1969. Growth of *Pseudomonas phaseolicola* in susceptible and in resistant bean plants. Ann. App. Biol. 63:103-116.

Otsuki, J., Shimomura, T., and Takebe, I. 1972. Tobacco mosaic virus multiplication and expression of the N gene in necrotic responding tobacco varieties. Virology 50:45-50.

Owen J.H., 1988. Role of abscisic acid in a C^{++} second messenger system. Physiol. Plant. 72:637-641.

Padgett, H.S., and Beachy, R.N. 1993. Analysis of tobacco mosaic virus strain capable of overcoming N gene-mediated resistance. Plant Cell. 5:577-586

Panopoulos, N.J., and Schroth, M.N. 1974. Role of flagellar motility in the invasion of bean leaves by *Pseudomonas phaseolicola*. Phytopathology 64:1389-1397.

Parish, C.L., Zaitlin, M., and Siegel, A. 1965. A study of necrotic lesion formation by tobacco mosaic virus, 2. Virology 26:413-418.

Pavlovkin, J., and Brinckmann, E. 1989. Increase of hydraulic conductivity during hypersensitive reaction induced by bacteria inoculation into *Nautilocalyx* leaves. Pages 589-590 in: Plant Membrane Transport: The Current Position. J. Dainty, I. DeMichelis, E. Marre, F. Rasi-Caldono, eds. Elsevier, Amsterdam.

Pavlovkin, J., Novacky, A., and Ullrich-Eberius, C.I. 1986. Membrane potential changes during bacteria-induced hypersensitive reaction. Physiol. Plant Pathol. 28:125-135.

Peever, T.L., and Higgins, V.J. 1989. Electrolyte leakage, lipoxygenase, and lipid peroxidation induced in tomato leaf tissue by specific and nonspecific elicitors from *Cladosporium fulvum*. Plant Physiol. 90:867-875

Pell, E.J., Lukezic, F.L., Levine, R.G., and Weissberger, W.C. 1977. Response of soybean foliage to reciprocal challenges by ozone and a hypersensitive-response-inducing pseudomonad. Phytopathology 67:1342-1345.

Peng, M., and Kuc, J. 1992. Peroxidase-generated hydrogen peroxide as a source of antifungal activity in vitro and on tobacco leaf disks. Phytopathology 82:696-699.

Pennel, C. 1986. The role of calcium in the control of peroxidase activity: Pages 155-168 in: Molecular and Physiological Aspects of Plant Peroxidases. H. Greppin, C. Penel and T. Gaspar, eds. Universite de Geneve, Switzerland.

Pennazio, S., and Roggero, P. 1990. Ethylene biosynthesis in soybean plants during the hypersensitive reaction to tobacco necrosis virus. Physiol. Mol. Plant Path. 36:121-128.

Pennazio, S., and Sapetti, C. 1982. Electrolyte leakage in relation to viral and abiotic stresses inducing necrosis in cowpea leaves. Biol. Plant. (Prague) 24:218-225.

Pfitzner, U.M., and Pfitzner, A.J.P. 1992. Expression of a viral avirulence gene in transgenic plants is sufficient to induce the hypersensitive defense reaction. Mol. Plant-Microb. Interact 5:318-321.

Phelps, R.H., and Sequeira, L. 1967. Synthesis of indoleacetic acid by a cell-free system from virulent and avirulent strains of *Pseudomonas solanacearum*. Phytopathology 57:1182-1190.

Pierce, M., and Essenberg, M. 1987. Localization of phytoalexins in fluorescent mesophyll cells isolated from bacterial blight-infected cotton cotyledons and separated from other cells by fluorescence-activated cell sorting. Physiol. Mol. Plant Pathol. 31:273-290.

Pierce, M.L., Essenberg, M., and Mort, A.J. 1993. A comparison of the quantities of exopolysaccharide produced by *Xanthomonas campestris* pv. *malvacearum* in susceptible and resistant cotton cotyledons during early stages of infection. Phytopathology 83:344-349.

Pike, S.M., and Novacky, A. 1988. Pathologically opened stomata: a mechanism for tissue desiccation in bacterial hypersensitivity? Page 233 in: Current Topics in Plant Biochemistry and Physiology, Vol. 7. D.D. Randall, D.G. Blevins, W.H. Campbell, J.M. Long, B.J. Rapp, eds. Interdisciplinary Plant Biochemistry and Physiology Program, University of Missouri, Columbia, MO.

Pike, S., Popham, P., Novacky, A., and Freeman, J. 1992. Intracellular pH alterations during bacterial HR: a confocal laser scanning microscopy study. (Abstr.) Phytopathology 82:1165.

Pike, S.M., Urbanek, H., and Novacky, A. 1991. Electrophysiological study of hypersensitive reaction to *Xanthomonas malvacearum* in resistant cotton (*Gossypium hirsutum* cv. Im216). (Abstr.) Phytopathology 81:1197.

Pinkas, Y., and Novacky, A. 1971. The differentiation between bacterial hypersensitive reaction and pathogenesis by the use of cycloheximide. (Abstr.) Phytopathology 61:906-907.

Politis, D.J., and Goodman, R.N. 1978. Localized cell wall appositions: Incompatibility response of tobacco leaf cells to *Pseudomonas pisi*. Phytopathology 68:309-316.

Poole, R.J. 1978. Energy coupling for membrane transport. Annu. Rev. Plant Physiol. 29:437-460.

Popham, P.L., and Novacky, A. 1991. Use of dimethyl sulfoxide to detect hydroxyl radical during bacteria-induced hypersensitive reaction. Plant Physiol. 96:1157-1160.

Popham, P., Pike, S., and Novacky, A. 1992. Electrical membrane potentials of cotton suspension-cultured cells exposed to hypersensitive reaction (HR)-inducing bacteria. (Abstr.) Phytopathology 82:1165.

Popham, P.L., Pike, S.M., Novacky, A., and Pallardy, S.G. 1993. Water relation alterations observed during bacterially induced hypersensitive reaction. Plant Physiol. in press.

Potts, J.R.M., Weklych, R., and Conn, R.E. 1974. The 4-hydroxylation of cinnamic acid by sorghum microsomes and the requirement for cytochrome P-450. J. Biol. Chem. 249:5019-5026.

Preisig, C.L., and Kuc, J. 1985. Arachidonic acid-related elicitors of the hypersensitive response in potato and enhancement of their activities by glucans from *Phytophthora infestans* (Mont.) de Bary. Arch. Biochem. Biophys. 236:379-389.

Preisig, C.L., and Kuc, J. 1987a. Inhibition by salicylhydroxamic acid, BW755C, eicosatetraenoic acid and disulfiram of hypersensitive resistance elicited by arachidonic acid and poly-l-lysine in potato tuber. Plant Physiol. 84:891-894.

Preisig, C.L., and Kuc, J. 1987b. Phytoalexins, elicitors, enhancers, suppressors, and other considerations in the regulation of *R-Gene* resistance to *Phytophthora infestans* in potato. Pages 203-221 in: Molecular Determinants of Plant Diseases. S. Nishimura, eds. Tokyo/Springer-Verlag, Berlin.

Preisig, C.L., and Kuc, J. 1988. Metabolism by potato tuber of arachidonic acid, an elicitor of hypersensitive resistance. Physiol. Mol. Plant Pathol. 32:77-88.

Prusky, D., Dinoor, A., and Jacoby, B. 1980. The sequence of death of haustoria and host cells during the hypersensitive reaction of oat to crown rust. Physiol. Plant Pathol. 17:33-40.

Pryor, N.A. 1976. The role of free radical reactions in biological systems. Pages 1-49 in: Free Radicals in Biology, Vol. 1. W.A. Pryor, ed. Academic Press, New York.

Pryor, T. 1987. The origin and structure of fungal disease resistance genes in plants. Trends Genet. 3:157-161.

Pueppke, S. G. 1984. Adsorption of bacteria to plant surfaces. Pages 215-261 in: Plant-Microbe Interaction, Vol. l. T. Kosuge and E. W. Nester, eds. Macmillan Publishing Co., New York.

Purohit, A.N., Tregunna, E.B., and Ragetli, H.W.J. 1975. CO_2 Effects on local-lesion production by tobacco mosaic virus and turnip mosaic virus. Virology 65:558-564.

Ragetli, H.W.J. 1967. Virus host interactions with emphasis on certain cytopathic-phenomena. Can. J. Bot. 45:1221-1234.

Rahme, L.G., Mindrinos, M.N., and Panopoulos, N.J. 1991. Genetic and transcriptional organization of the *hrp* cluster of *Pseudomonas syringae* pv. *phaseolicola*. J. Bacteriol. 173:575-586.

Raskin, I. 1992. Role of salicylic acid in plants. Annu. Rev. Plant Physiol. Plant Mol. Biol. 43:439-463.

Rasmussen, H., and Barrett, P.Q. 1984. Calcium messenger system: an integrated view. Physiol. Rev. 64:938-984.

Rasmussen, J.B., Hammerschmidt, R., and Zook, M. 1991. Systemic induction of salicylic acid accumulation in cucumber after inoculation with *Pseudomonas syringae* pv *syringae*. Plant Physiol. 97: 1342-1347.

Rich. P.R., and Lamb, J.C. 1977. Biophysical and enzymological studies upon the interaction of *trans*-cinnamic acid with higher plant microsomal cytochromes *P*-450. Eur. J. Biochem. 72:353-360.

Ricker, K.E., and Bostock, R.M. 1992. Evidence for release of the fungal elicitor arachidonic acid and its metabolites from sporangia of *Phytophthora infestans* during infection of potato. Physiol. Mol. Plant Pathol. 41:61-72.

Riggle, J.H., and Klos, E.J. 1972. Relationship of *Erwinia herbicola* to *Erwinia amylovora*. Can. J. Bot. 50:1077-1083.

Riker, A.J. 1923. Some relations of the crown gall organism to its host tissue. J. Agric. Res. 25:119-132.

Roberts, D.P., Denny, T.P., and Schell, M.A. 1988. Cloning of the egl gene of *Pseudomonas solanacearum* and analysis of its role in phytopathogenicity. J. Bacteriol. 170:1445-1451.

Roberts, E., and Kolattukudy, P.E. 1989. Molecular cloning, nucleotide sequence, and abscisic acid induction of a suberization-associated highly anionic peroxidase. Mol. & Gen. Genet. 217:223-232.

Roberts, E., Kutchan, T., and Kolattukudy, P.E. 1988. Cloning and sequencing of cDNA for a highly anionic peroxidase from potato and the induction of its mRNA in suberizing potato tubers and tomato fruits. Plant Mol. Biol. 11:15-26.

Roche, P., Debelle, F., Maillet, F., Lerouge, P., Faucher, C., Truchet, G., Denarie, J., and Prome, J. 1991. Molecular basis of symbiotic host specificity in *Rhizobium meliloti*: nodH and nodPQ genes encode the sulfation of lipo-oligosaccharide signals. Cell 67: 1131-1143.

Roebuck, P., Sexton, R., and Mansfield, J.W. 1978. Ultrastructural observations on the development of the hypersensitive reaction in leaves of *Phaseolus vulgaris* cv. Red Mexican inoculated with *Pseudomonas phaseolicola* (race 1). Physiol. Plant Pathol. 12:151-157.

Rogers, K.R., Albert, F., and Anderson, A.J. 1988. Lipid peroxidation is a consequence of elicitor activity. Plant Physiol. 86:547-553.

Romberger, J.A., and Norton, G. 1961. Changing respiratory pathways in potato slices. Plant Physiol. 36:20-29.

Rona, J.P., Pitman, G.M., Luttge, U., and Ball, E. 1980. Electrochemical data on compartimentation into wall, cytoplasm and vacuole of leaf cells in the CAM genus *Kalanchoe*. J. Memb. Biol. 57:25-35.

Rudolph, K. 1975. Models of interaction between higher plants and bacteria. Pages 109-129 in: Specificity in Plant Diseases. R.K.S. Wood and A. Graniti, ed. Plenum Press. New York and London.

Rudolph, K. 1978. A host specific principle from *Pseudomonas phaseolicola* (Burkh.) Dowson, inducing water-soaking in bean leaves. Phytopathol. Z. 93:218-226.

Rudolph, K. 1984. Multiplication of *Pseudomonas syringae* pv. *phaseolicola* "in planta". I. Relation between bacterial concentration and water-congestion in different bean cultivars and plant species. Phytopathol. Z. 111:349-362.

Rudolph, K.W.E., Gross, M., Neugebauer, M., Hokawat, S., Zachowski, A., Wydra, K., and Klement, Z. 1989. Extracellular polysaccharides as determinants of leaf spot diseases caused by pseudomonads and xanthomonads. Pages 177-218 in: Phytotoxins and Plant Pathogenesis. A. Graniti, R. Durbin, and A. Ballio, eds. Springer-Verlag, Berlin, Heidelberg.

Ruzicska, P., Gombos, Z., and Farkas, G.L. 1983. Modification of the fatty acid composition of phospholipids during the hypersensitive reaction in tobacco. Virology 128:60-64.

Sakai, R., and Tomiyama, K. 1964. Relation between physiological factors related to phenols and varietal resistance of potato plant to late-blight. I. Field observation. Ann. Phytopathol. Soc. Jpn. 29:33-38.

Sakai, R., Tomiyama, K., Ishizaka, N., and Sato, N. 1967. Phenol metabolism in relation to disease resistance of potato tubers. 3. Phenol metabolism in tissue neighbouring the necrogenous infection. Ann. Phytopathol. Soc. Jpn. 33:216-222.

Sakai, R., Tomiyama, K., and Takemori, T. 1964. Relation between physiological factors related to phenols and varietal resistance of potato plant to late blight. II. Experiments with the potato tubers. Ann. Phytopathol. Soc. Jpn. 29:120-127.

Sakai, S., Doke, N., and Tomiyama, K. 1982. Relation between necrosis and rishitin accumulation in potato tuber slices treated with hyphal wall components of *Phytophthora infestans*. Ann. Phytopathol. Soc. Jpn. 48:238-240.

Sakai, S., Tomiyama, K., and Doke, N. 1979. Synthesis of a sesquiterpenoid phytoalexin rishitin in non-infected tissue from various parts of potato plants immediately after slicing. Ann. Phytopathol. Soc. Jpn. 45:705-711.

Sakai, S., Tomiyama, K., and Doke, N. 1981. Effect of hyphal wall component of *Phytophthora infestans* on activity of rishitin synthetic pathway in a potato cell exposed directly to the component. Ann. Phytopathol. Soc. Jpn. 47:255-257.

Sakuma, T., and Tomiyama, K. 1967. The role of phenolic compounds in the resistance of potato tuber tissue to infection by *Phytophthora infestans*. Ann. Phytopathol. Soc. Jpn. 33:48-58.

Salin, M.L., and Bridges, S.M. 1981. Chemiluminescence in wounded root tissue. Plant Physiol. 67:43-46.

Salzwedel, J.L., Daub, M.E., and Huang, J. 1989. Effects of singlet oxygen quenchers and pH on the bacterially induced hypersensitive reaction in tobacco suspension cell cultures. Plant Physiol. 90:25-28.

Samborski, D.J., Kim, W.K., Rohringer, R., Howes, N.K., and Baker, R.J. 1977. Histological studies on host-cell necrosis conditioned by the *SR6* gene for resistance in wheat to stem rust. Can. J. Bot. 55:1445-1452.

Samuel, G. 1931. Some experiments on inoculating methods with plant viruses, and on local lesions. Ann. Biol. (Ludhiana) 18:494-507.

Sasser, M. 1978. Involvement of bacterial protein synthesis in induction of the hypersensitive reaction in tobacco. Phytopathology 68:361-363.

Sasser, M. 1982. Inhibition by antibacterial compounds of the hypersensitive reaction induced by *Pseudomonas pisi* in tobacco. Phytopathology 72:1513-1517.

Sato, N., Kitazawa, K., and Tomiyama, K. 1971. The role of rishitin in localizing the invading hyphae of *Phytophthora infestans* in infection sites at the cut surface of potato tubers. Physiol. Plant Pathol. 1:289-295.

Sato, N., and Tomiyama, K. 1969. Localized accumulation of rishitin in the potato-tuber tissue infected by an incompatible race of *Phytophthora infestans*. Ann. Phytopathol. Soc. Jpn. 35:202-217.

Sato, N., and Tomiyama, K. 1976a. Relation between rishitin accumulation and degree of resistance of potato-tuber tissue to infection by an incompatible race of *Phytophthora infestans*. Ann. Phytopathol. Soc. Jpn. 42:431-435.

Sato, N., and Tomiyama, K. 1976b. Relationship between inhibition of intracellular hyphal growth to *Phytophthora infestans* and rishitin concentration in infected potato cells. Ann. Phytopathol. Soc. Jpn. 43:598-600.

Schaad, N.W., Heskett, M.G., Gardener, J.M., and Kado, C.I. 1973. Influence of inoculum dosage, time after wounding, and season on infection of Persian walnut trees by *Erwinia rubrifaciens*. Phytopathology 63:327-329.

Schlumbaum, A., Mauch, F., Vogeli, U., and Boller, T. 1986. Plant chitinases are potent inhibitors of fungal growth. Nature 324:365-367.

Schmidt, W.E., and Ebel, J. 1987. Specific binding of a fungal glucan phytoalexin elicitor to membrane fractions from soybean *Glycine max*. Proc. Natl. Acad. Sci. USA. 84:4117-4121.

Scholtens-Toma, I.M.J., and deWit, P.J.G.M. 1988. Purification and primary structure of a necrosis-inducing peptide from the apoplastic fluids of tomato infected with *Cladosporium fulvum* (syn. *Fulvia fulva*) Physiol. Mol. Plant Pathol. 33:59-67.

Schroeder, J.I., and Hagiwara S. 1989. Cytosolic calcium regulates ion channels in the plasma membrane of *Vicia faba*. Nature 312:361-362.

Schulte, R., and Bonas, U. 1992. Expression of the *Xanthomonas campestris* pv. *vesicatoria* *hrp* gene cluster, which determines pathogenicity and hypersensitivity on pepper and tomato, is plant inducible. J. Bacteriol. 174:815-823.

Schwacke, R., and Hager, A. 1992. Fungal elicitors induce a transient release of active oxygen species from cultured spruce cells that is dependent on Ca^{2+} and protein-kinase activity. Planta 187:136-141

Scott, I.M., Martin, G.C., Horgan, R., and Heald, J.K. 1982. Mass spectrometric measurement of zeatin glycoside levels in *Vinca rosea* L. crown gall tissue. Planta 154:273-276.

Segal, A.W., and Abo. A. 1993. The biochemical basis of the NADPH oxidase of phagocytes. Trends Biochem. Sci. 10:43-47.

Sequeira, L. 1975. Induction and suppression of the hypersensitive reaction caused by phytopathogenic bacteria: specific and non-specific components. Pages 289-306 in: Specifity in Plant Diseases. R.K.S. Wood and A. Graniti, eds. Plenum Press, New York.

Sequeira, L. 1983. Mechanisms of induced resistance in plants, fungi, viruses, bacteria. Annu. Rev. Microbiol. 37: 51-79.

Sequeira, L. 1985. Surface components involved in bacterial pathogen-plant host recognition. J. Cell Sci. Suppl. 2: 301-316.

Sequeira, L., and Ainslie, V. 1969. Bacteria cell-free preparations that induce or prevent the hypersensitive reaction in tobacco. (Abstr.) Proceedings of the XI th International Botanical Congress, page 195.

Sequeira, L., Gaard, G., and de Zoeten G.A. 1977. Attachment of bacteria to host cell walls: its relation to mechanisms of induced resistance. Physiol. Plant Pathol. 10:43-50.

Sequeira, L. and Hill, L. M. 1974. Induced resistance in tobacco leaves: the growth of *Pseudomonas solanacearum* in protected tissues. Physiol. Plant Pathol. 4:447-455.

Serrano, R. 1989. Structure and function of plasma membrane ATPase. Annu. Rev. Plant Physiol. Plant Mol. Biol. 40:61-94.

Sharp, E.L., and Emge, R.G. 1958. A "tissue transplant" technique for obtaining abundant sporulation of races of *Puccinia graminis* var. *tritici* on resistant varieties. Phytopathology 48:696-697.

Shimomura, T. 1971. Necrosis and localization of infection in local lesion hosts. Phytopathol. Z. 70:185-196.

Shimony, C., and Friend, J. 1975. Ultrastructure of the interaction between *Phytophthora infestans* and leaves of two cultivars of potato (*Solanum tuberosum L.*) orion and majestic. New Phytol. 74:59-65.

Shintaku, M.H., and Patil, S.S. 1986. Isolation of an avirulence determinant from *Pseudomonas syringae* pv. *phaseolicola*. (Abstr.) Third International Symposium on the Molecular Genetics of Plant-Microbe Interactions, Montreal, Canada. page 38.

Shiraishi, R., Yamaoka, N., and Kunoh, H. 1989. Association between increased phenylalanine ammonialyase activity and cinnamic acid synthesis and the induction of temporary inaccessibility caused by *Erysiphe graminis* primary germ tube penetration of the barley leaf. Physiol. Mol. Plant Pathol. 34:75-83.

Siegel, A. 1966. The first stages of infection. Pages 5-18 in: Viruses of Plants. A. Beemster and J. Dijkstray, eds. No. Holland Publ. Co., Amsterdam.

Sigee, D.C., and Al-Issa, A.N. 1982. The hypersensitive reaction in tobacco leaf tissue infiltrated with *Pseudomonas pisi*. 2. Changes in the population of viable, actively metabolic and total bacteria. Phytopathol. Z. 105:71-86.

Silverman, W. 1959. The effect of variations in temperature on the necrosis associated with infection type 2 uredia of the wheat stem rust fungus. Phytopathol. 49:827-830.

Simon, E. W. 1974. Phospholipids and plant membrane permeability. New Phytol. 73:377-420.

Siwecki, G., and Brown, O.R. 1990. Overproduction of superoxide dismutase does not protect *Escherichia coli* from stringency-induced growth inhibition by 1mM paraquat. Biochem. Inter. 20:191-200.

Skipp, R.A., Harder, D.E., and Samborski, D.J. 1974. Electron microscopy studies on infection of resistant (*Sr6* gene) and susceptible near-isogenic wheat lines by *Puccinia graminis* f. sp. *tritici*. Can. J. Bot. 52:2615-2620.

Smith, F.A., and Raven, J.A. 1979. Intracellular pH and its regulation. Annu. Rev. Plant Physiol. 30: 289-311.

Smith, J.A., Hammerschmidt, R., and Fulbight, D.W. 1991. Rapid induction of systemic resistance in cucumber by *Pseudomonas syringae* pv. *syringae*. Physiol. Mol. Plant Pathol. 38:223-235.

Solymosy, F., and Farkas, G.L. 1963. Metabolic characteristics at the enzymatic level of tobacco tissues exhibiting localized acquired resistance to viral infection. Virology. 21:210-221.

Somlyai, G., Hevesi, M. Banfalvi, Z., Klement, Z., and Kondorosi, A. 1986. Isolation and characterization of non-pathogenic and reduced virulence mutants of *Pseudomonas syringae* pv. *phaseolicola* induced by Tn*5* transposon insertions. Physiol. Mol. Plant Pathol. 29:369-380.

Somlyai, G., Solt, A., Hevesi, M., El-Kady, S., Klement, Z., and Kari, C. 1988. The relationship between the growth rate of *Pseudomonas syringae* pathovars and the hypersensitive reaction in tobacco. Physiol. Mol. Plant Pathol. 33:473-482.

Spanswick, R.M. 1981. Electrogenic ion pumps. Annu. Rev. Plant Physiol. 32:267-289.

Spielman, L.J., McMaster, B.J., and Fry, W.E. 1989. Dominance and recessiveness at loci for virulence against potato and tomato in *Phytophthora infestans*. Theor. Appl. Genet. 77:832-838.

Srivastava, G.P., and Van Huystee, R.B. 1977. Interaction among phenolics and peroxidase enzymes. Bot. Gaz. 138:452-464.

Stab, M.R., and Ebel, J. 1987. Effects of Ca^{2+} on phytoalexin induction by fungal elicitor in soybean cells. Arch. Biochem. Biophys. 257:416-423.

Stachel, S.E., Messens, E., Van Montague, M., and Zambryski, P. 1985. Identification of the signal molecules produced by wounded plant cells that activate T-DNA transfer in *Agrobacterium tumefaciens*. Nature 318:624-629.

Stadelmann, J. 1969. Permeability of the plant cell. Annu. Rev. Plant Physiol. 20:585-605.

Stakman, E.C. 1915. Relation between *Puccinia graminis* and plants highly resistant to its attack. J. Agric. Res. 4:193-299.

Stall, R.E., and Cook, A.A. 1973. Hypersensitivity as a defense mechanism against natural infection. Abstract No. 0586 In: Abstracts of Papers 2nd Int. Congr. Plant Pathol., 5-12 Sept. 1973, Minneapolis, Minnesota (unpaged).

Stall, R.E., and Cook, A.A. 1979. Evidence that bacterial contact with the plant cell is necessary for the hypersensitive reaction but not the susceptible reaction. Physiol. Plant Pathol. 14:77-84.

Stall, R.E., Hall, C.B., Cook, A.A. 1972. Relationship of ammonia to necrosis of pepper leaf tissue during colonization by *Xanthomonas vesicatoria*. Phytopathology 62:882-886.

Stall, R.E., Loschke, D.C., and Jones, J.B. 1986. Linkage of copper resistance and avirulence loci on a self-transmissible plasmid in *Xanthomonas campestris* pv. *vesicatoria*. Phytopathol. 76:240-243.

Stall, R.E., and Minsavage, G.V. 1990. The use of *hrp* genes to identify opportunistic xanthomonads. Pages 369-374 in: Proceedings of the 7th International Conference on Plant Pathogenic Bacteria, Budapest, Hungary. June 11-16, 1989:Plant Pathogenic Bacteria. Z. Klement, ed. Akademiai Kiado, Budapest.

Standen, N.B., Gray, P.T.A., and Whitaker, M.J. 1987. Microelectrode Techniques. The Company of Biologists Limited, Cambridge.

Staskawicz, B.J., Dahlbeck, D., and Keen, N.T. 1984. Cloned avirulence gene of *Pseudomonas syringae* pv. *glycinea* determines race-specific incompatibility on *Glycine max*. (L.) Merr. Proc. Natl. Acad. Sci. USA. 81:6024-6028.

Staskawicz, B., Dahlbeck, D., Keen, N., and Napoli, C. 1987. Molecular characterization of cloned avirulence genes from race 0 and race 1 of *Pseudomonas syringae* pv. *glycinea*. J. Bacteriol. 169:5789-5794.

Steere, R.L. 1955. Concepts and problems concerning the assay of plant viruses. Phytopathology 45:196-208.

Steinberger, E.M., and Beer, S.V. 1988. Creation and complementation of pathogenicity mutants of *Erwinia amylovora*. Mol. Plant-Microbe Interact. 1:135-144.

Stemmer, W.P., and Sequeira, L. 1984. Possible role of pili in attachment of *Pseudomonas solanacearum* to plant cell walls. Page 99 in: 2nd Int. Symp. Molecular Genetics of Plant-Bacteria Interactions, Abstracts. Ithaca, N.Y.

Stermer, B.A., and Bostock, R.M. 1987. Involvement of 3-hydroxy-3-methylglutaryl coenzyme A reductase in the regulation of sesquiterpenoid phytoalexin synthesis in potato. Plant Physiol. 84:404-408.

Stermer, B.A., and Bostock, R.M. 1989. Rapid changes in protein synthesis after application of arachidonic acid to potato tuber tissue. Physiol. Mol. Plant Pathol. 35:347-356.

Stewart, P.A. 1981. How to understand acid-base: a quantitative acid-base primer. Elsevier North Holland, Inc.

Sticher, L., Pennel, C., and Greppin, H. 1981. Calcium requirement for the secretion of peroxidases by plant cell suspensions. J. Cell Sci. 48:345-353.

Sulzinski, M.A., and Zaitlin, M. 1982. Tobacco mosaic virus replication in resistant and susceptible plants: In some resistant species virus is confined to a small number of initially infected cells. Virology 121:12-19.

Sutherland, I.W. 1988. Bacterial surface polysaccharides: structure and function. Int. Rev. Cytol. 113:187-231.

Svingen, B.A., Buege, J.A., O'Neal, F., and Aust, S.D. 1979. The mechanism of NADPH-dependent lipid peroxidation. The propagation of lipid peroxidation. J. Biol. Chem. 254:5892-5899.

Swarup, S., DeFeyter, R., Brlansky, R.H., and Gabriel, D.W. 1991. A pathogenicity locus from *Xanthomonas citri* enables strains from several pathovars of *X. campestris* to elicit cankerlike lesions on citrus. Phytopathology 81:802-809.

Swarup, S., Yang, Y., Kingsley, M.T., and Gabriel, D.W. 1992. An *Xanthomonas citri* pathogenicity gene, *pthA*, pleiotropically encodes gratuitous avirulence on nonhosts. Mol. Plant-Microbe Interact. 5:204-213.

Takahashi, T. 1973. Studies on viral pathogenesis in plant hosts IV. Comparison of early process of tobacco mosaic virus infection in the leaves of 'Samsun NN' and "Samsun" tobacco plants. Phytopathol. Z. 77:157-168.

Takamatsu, N., Ishikawa, M., Meshi, T., and Okada, Y. 1987. Expression of bacterial chloramphenicol acetyltransferase gene in tobacco plants mediated by TMV-RNA. Eur. Mol. Biol. Organ. J. 6:307-311.

Taylor, A.E., and Townsley, M.I. 1986. Assessment of oxygen radical tissue damage. Pages 19-38 in: Physiology of Oxygen Radicals. A.E. Taylor, S. Matalon, and P. Ward, eds. Amer. Physiol. Soc. Bethesda, MD.

Templeton, M.D., and Lamb, C.J. 1988. Elicitors and defense gene activation. Plant, Cell Environ. 11:395-401.

Tepfer, M., and Taylor, I.T.P. 1981. The permeability of plant cell walls as measured by gel filtration chromatography. Science 213:761-763.

Thatcher, F.S. 1942. Further studies on osmotic and permeability relations in parasitism. Can. J. Res. Sect. C. Bot. Sci. 20:283-311.

Theologis, A., and Laties, G.G. 1981. Wound-induced membrane lipid breakdown in potato tuber. Plant Physiol. 68:53-58.

Thompson, J.E., Legge, R.L., and Barber, R.F. 1987. The role of free radicals in senescence and wounding. New Phytol. 105:317-344.

Tiburzy, R., and Reisener, H.J. 1990. Resistance of wheat to *Puccinia graminis* f. sp. *tritici*: Association of the hypersensitive reaction with the cellular accumulation of lignin-like material and callose. Physiol. Mol. Plant Pathol. 36:109-120.

Tjamos, E.C., and Kuc, J.A. 1982. Inhibition of steroid glycoalkaloid accumulation by arachidonic and eicosapentaenoic acids in potato. Science, 217:542-543.

Tomiyama, K. 1955. Cytological studies on resistance of potato plants to *Phytophthora infestans*. II. The death of the intracellular hyphae in the hypersensitive cell. Japan. Research Bulletin of the Hokkaido Nat. Agric. Expt. Sta., Sapporo, Japan. Pages 149-154.

Tomiyama, K. 1956a. Cytological studies on resistance of potato plants to *Phytophthora infestans*. III. The time required for the browning of midrib cell of potato plants infected by *P. infestans*. Research Bulletin of the Hokkaido Nat. Agric. Expt. Sta., Sapporo, Japan. Pages 165-169.

Tomiyama, K. 1956b. Cell physiological studies on the resistance of potato plant to *Phytophthora infestans*. IV. On the movements of cytoplasm of the host cell induced by the invasion of *Phytophthora infestans*. Research bulletin of the Hokkaido Nat. Agri. Expt. Sta., Sapporo, Japan. Pages 54-62.

Tomiyama, K. 1957a. Cell physiological studies on the resistance of potato plants to *Phytophthora infestans*. V. Effect of 2,4-dinitrophenol upon the hypersensitive reaction of potato plant cell to infection by *P. infestans*. Ann. Phytopathol. Soc. Jpn. 22:75-78.

Tomiyama, K. 1957b. Cell physiological studies on resistance of potato plants to *Phytophthora infestans*. VII. Growth of intracellular hyphae of *Phytophthora infestans* in the living potato plant cells which are resistant or susceptible to the infection of them. Research Bulletin of the Hokkaido Nat. Agri. Expt. Sta. Sapporo, Japan. Pages 129-134.

Tomiyama, K. 1960. Some factors affecting the death of hypersensitive potato plant cells infected by *Phytophthora infestans*. Phytopathol. Z. 39:134-148.

Tomiyama, K. 1966. Double infection by an incompatible race of *Phytophthora infestans* of a potato plant cell which has previously been infected by a compatible race. Ann. Phytopathol. Soc. Jpn. 32:181-185.

Tomiyama, K. 1967. Further observation on the time requirement for hypersensitive cell death of potatoes infected by *Phytophthora infestans* and its relation to metabolic activity. Phytopathol. Z. 58:367-378.

Tomiyama, K. 1971. Cytological and biochemical studies of the hypersensitive reaction of potato cells to *Phytophthora infestans*. Pages 387-397 in: Morphological and Biochemical Events in Plant-Parasite Interaction. S. Akai and S. Ouchi, eds. Phytopathol. Soc. of Japan.

Tomiyama, K. 1989. Biochemical aspects of potato late blight with respect to compatibility and incompatibility reactions. Report of the Planning Conference on Fungal Diseases of the Potato. Held at CIP, Lima, Peru.

Tomiyama, K., and Ishizaka, N. 1967. Phenol metabolism in relation to disease resistance of potato tuber. II. Exponential equation relating levels of phenols or phenolic enzymes to distance from locus of injury of plant tissue. Plant Cell Physiol. 8:217-220.

Tomiyama, K., Ishizaka, N., Sato, N., Masamune, T., and Katsui, N. 1968. "Rishitin" A phytoalexin-like substance. Its role in the defence reaction of potato tuber to infection. Phytopathol. Soc. Japan. Tokyo. Pages 287-292.

Tomiyama, K., Lee, H.S., and Doke, N. 1974. Effect of hyphal homogenate of *Phytophthora infestans* on the potato-tuber protoplasts. Ann. Phytopathol. Soc. of Jpn. 40:70-72.

Tomiyama, K., Okamoto, H., and Katou, K. 1983. Effect of infection by *Phytophthora infestans* on the membrane potential of potato cells. Physiol. Plant Pathol. 22:233-243.

Tomiyama, K., Okamoto, H., and Katou, K. 1987. Membrane potential change induced by infection by *Phytophthora infestans* of potato cells. Ann. Phytopathol. Soc. Jpn. 53:310-332.

Tomiyama, K., and Okamoto, H. 1989. Effect of pre-inoculational heat-treatment on the membrane potential change of potato cell induced by infection with *Phytophthora infestans*. Physiol. Mol. Plant Pathol. 35:191-201.

Tomiyama, K., Sakai, R., Otani, Y., and Takemorei, T. 1967. Phenol metabolism in relation to disease resistance of potato tuber. I. Activities of phenol oxidase and some enzymes related to glycolysis as affected by chlorogenic acid treatment. Plant Cell Physiol. 8:1-13.

Tomiyama, K., Sakai, R., Takase, N., and Takakuwa, M. 1956. Physiological studies on the defence reaction of potato plant to the infection by *Phytophthora infestans*. IV. The influence of pre-infectional ethanol narcosis upon the physiological reaction of potato tuber to the infection by *P. infestans*. (part 2). Ann. Phytopathol. Soc. Jpn. 21:153-158.

Tomiyama, K., Sakuma, T., Ishizaka, N., Sato, N., Katsue, N., Takasugi, M., and Masamune, Y. 1968. A new antifungal substance isolated from resistant potato tuber tissue infected by pathogens. Phytopathology 58:115-116.

Tomiyama, K., and Stahmann, M.A. 1964. Alteration of oxidative enzymes in potato tuber tissue by infection with *Phytophthora infestans*. Plant Physiol. 39:483-490.

Tomiyama, K., Takakuwa, M., Takase, N., and Sakai, R. 1956. Physiological studies on the defense reaction of potato plant to the infection of *Phytophthora infestans*. III. The influence of pre-infectional ethanol narcosis upon the physiological reaction of potato tuber to the infection of *P. infestans*. (part I). Ann. Phytopathol. Soc. Jpn. 21:17-22.

Tomiyama, K., Takakuwa, M., and Takase, N. 1958. The metabolic activity in healthy tissue neighboring the infected cells in relation to resistance to *Phytophthora infestans* (Mont.) DeBary in potatoes. Phytopathol. Z. 31:237-250.

Tomiyama, K., Takakuwa, M., Takase, N., and Sakai, R. 1959. Alteration of oxidative metabolism in a potato tuber cell invaded by *Phytophthora infestans* and in the neighbouring tissues. Phytopathol. Z. 37:113-144.

Tomiyama, K., Takase, N., Sakai, R., and Takakuwa, M. 1955. Physiological studies on the defence reaction of potato plant to the infection by *Phytophthora infestans*. II Changes in the physiology of potato tuber induced by the infection of the different strains of *P. infestans*. Ann. Phytopathol. Soc. Jpn. 20:59-64.

Tomiyama, K., Takase, N., Sakai, R., and Takakuwa, M. 1956. Physiological studies on the defense reaction of potato plant to the infection of *Phytophthora infestans*. I Changes in the physiology of potato tuber induced by the infection of *P. infestans* and their varietal differences. Research Bulletin of the Hokkaido National Agricultural Experiment Station, Sapporo, Japan. 71:32-50.

Tukey, H.B.J. 1970. The leaching of substances from plants. Annu. Rev. Plant Physiol. 21:305-324.

Turner, J.G. 1976. The non-electrolyte permeability of tobacco cell membranes in tissues inoculated with *Pseudomonas pisi* and *Pseudomonas tabaci*. Ph.D. Thesis, University of Missouri-Columbia, Pages 1-154.

Turner, J.G., and Novacky, A. 1974. The quantitative relation between plant and bacterial cells involved in the hypersensitive reaction. Phytopathology 64:885-890.

Turner, J.G., and Novacky, A. 1976. Effect of *Pseudomonas tabaci* and *P. pisi* on permeability of tobacco protoplasts to nonelectrolytes. Proc. Am. Phytopathol. Soc. 3:260-261.

Ullrich, W.R., Kunz, G., Stiller, A., and Novacky, A.J. 1993. Role of ammonium accumulation in bacteria-induced hypersensitive and compatible reactions of tobacco and cotton plants. Physiol. Plant. 89:644-652.

Ullrich, W.R., Novacky, A., and Kunz, G. 1984. The role of ammonium in bacterial plant diseases: accumulation and the effect on the electrical membrane potential in cotton and tobacco. Pages 514-515 in: Membrane Transport in Plants. W.J. Cram, K. Janacek, R. Rybova and K. Sigler, eds. Academia, Prague.

Ullrich, C.I., and Novacky, A. 1992. Recent aspects of ion-induced pH changes. Pages 231-248 in: Current Topics in Plant Biochemistry and Physiology. Vol. 11. D.D. Randall, R.E. Sharp, A.J. Novacky, and D.G. Blevins, eds. Interdisciplinary Plant Biochemistry and Physiology Program, University of Missouri, Columbia, MO.

Ullrich-Eberius, C.I., Pavlovkin, J. Schindel, J., Fischer, K., and Novacky, A. 1989. Changes in plasmalemma functions induced by phytopathogenic bacteria. Pages 323-332 in: Plasma Membrane Oxidoreductases in Control of Animal and Plant Growth. F.L. Crane, D.J. Morre and H. Low, eds. Plenum Press, New York and London.

Van den Ackerveken, A.F.J.M., Van Kan, J.A.L., and de Wit, P.J.G.M. 1991. Molecular evidence supporting the gene-for-gene hypothesis in *Cladosporium fulvum*-tomato interaction. Plant J. 2:359-366.

Van den Ackervekem. J.M., Vossen, P., and de Wit, P.J. 1993. The *avr9* race-specific elicitor of *Cladosporium fulvum* is processed by endogenous and plant proteases. Plant Physiol. 103:91-96.

Vanneste, J.L., Bauer, D.W., Beer, S.V., and Norelli, J.L. 1987. The cloning and restriction map of the recA-like gene of *Erwinia amylovora*. Pages 173-174 in: Fire blight Proceedings of an International Workshop, June 22-26, 1986, S.V. Beer, ed. Ithaca, New York.

Van Kan, J.A.L., Van Ackerveken, A.F.J.M., and de Wit, P.J.G.M. 1991. Cloning and characterization of avirulence gene *avr* 9 of the fungal pathogen *Cladosporium fulvum* causal agent of tomato leaf mold. Mol. Plant Microb. Interact. 4:51-59.

Van Loon, L.C. 1975. Polyacrylamide disc electrophoresis of soluble leaf proteins from *Nicotiana tabarum* var Samsun and Samsun NN. IV. Similarity of quantitative changes of specific proteins after infection with different viruses and their relationship to acquired resistance. Virology 67:566-575.

Van Loon, L.C. 1986. The significance of changes in peroxidase in diseased plants. Pages 405-415 in: Molecular and Physiological Aspects of Plant Peroxidases. H. Greppin, C. Penel, and T. Gaspar, eds. Univ. of Geneva, Switzerland.

Van Loon, L.C. 1987. Disease induction by plant viruses. Adv. Virus Res. 33:205-255.

Varns, J., Currier, W.W., and Kuc, J. 1971. Specificity of rishitin and phytuberin accumulation by potato. Phytopathology. 61:174-177.

Vaughn, S.F., and Lulai, E.C. 1992. Further evidence that lipoxygenase activity is required for arachidonic acid-elicited hypersensitivity in potato callus cultures. Plant Sci. 84:91-98.

Vera-Estrella, R., Blumwald, E., and Higgins, V.J. 1992. Effect of specific elicitors of *Cladosporium fulvum* on tomato suspension cells. Plant Physiol. 99:1208-1215.

Vickery, M.L., and Vickery, B. 1981. Secondary Plant Metabolism. University Park Press, Baltimore.

Walker, J.C. and Patel, P.N. 1964. Inheritance of resistance to halo blight of bean. Phytopathology 54:952-954.

Ward, H.M. 1902. On the relations between host and parasite in the bromes and their brown rust, *Puccinia dispersa* (Erikss.). Ann. Bot. (Lond.) 16:233-315.

Wehtje, G., Littlefield, L.J., and Zimmer, D.E. 1979. Ultrastructure of compatible and incompatible reactions of sunflower to *Plasmopara halstedii*. Can. J. Bot. 57:315-323.

Wei, Z-M., Laby, R.J., Zumoff, C.H., Bauer, D.W., He, S.Y., Collmer, A., and Beer, S.V. 1992a. Harpin, elicitor of the hypersensitive response produced by the plant pathogen *Erwinia amylovora*. Science 257:85-88.

Wei, Z-M., Sneath, B.J., and Beer, S.V. 1992b. Expression of *Erwinia amylovora hrp* genes in response to environmental stimuli. J. Bacteriol. 174:1875-1882.

Weintraub, M., Kemp, W.G., and Ragetli, H.W.J. 1960. Studies on the metabolism of leaves with localized virus infections. I. Oxygen uptake. Can. J. Microbiol. 6:407-415.

Weintraub, M., Kemp, W.G., and Ragetli, H.W.J. 1961. Some observations on hypersensitivity to plant viruses. Phytopathology 51:290-293.

Weintraub, M., and Kemp, W.G. 1961. Protection with carnation mosaic virus in *Dianthus barbatus*. Virology 13:256-257.

Weintraub, M., Kemp, W.G., and Ragetli, H.W.J. 1963. Conditions for systemic invasion by virus of a hypersensitive host. Phytopathology 53:618.

Weintraub, M., and Ragetli, H.W.J. 1961. Cell wall composition of leaves with a localized virus infection. Phytopathology 51:215-219.

Weintraub, M., and Ragetli, H.W.J. 1964a. Studies on the metabolism of leaves with localized virus infections. Can. J. Bot. 42:533-540.

Weintraub, M., and Ragetli, H.W.J. 1964b. An electron microscope study of tobacco mosaic virus lesions in *Nicotiana glutinosa* L.J. Cell Biol. 23:499-509.

Weintraub, M., and Ragetli, H.W.J. 1964c. Electron microscopy of chloroplast fragments in local lesions of TMV-infected *Nicotiana glutinosa*. Phytopathology. 54:870-871.

Weintraub, M., Ragetli, H.W.J., and Dwurazna, M.M. 1964. Studies on the metabolism of leaves with localized virus infections. Mitochondrial activity in TMV-infected *Nicotiana glutinosa* L. Can. J. Bot. 42:541-545.

Weintraub, M., Ragetli, H.W.J., and John, V.T. 1967. Some conditions affecting the intracellular arrangement and concentration of tobacco mosaic virus particles in local lesions. J. Cell Biol. 35:183-192.

Weintraub, M., and Ragetli, H.W.J. 1971. Some characteristics of a virus from Virginia crab apple. Phytopathology 61:431-432.

Weintraub, M., Ragetli, H.W.J., and Lo, E. 1972. Mitochondrial content and respiration in leaves with localized virus infections. Virology 50:841-850.

Weintraub, M., and Schroeder, B. 1979. Cytochrome oxidase activity in hypertrophied mitochondria of virus-infected leaf cells. Phytomorphology 29:273-285.

Wenham, R.J., Fraser, R.W.W., and Snow, A. 1985. Tobacco mosaic virus-induced increase in abscisic acid concentration in tobacco leaves: intracellular location and relationship to symptom severity and extent of virus multiplication. Physiol. Plant Pathol. 26:379-387.

Weststeijn, E.A. 1978. Permeability changes in the hypersensitive reaction of *Nicotiana tabacum* cv. Xanthi nc. after infection with tobacco mosaic virus. Physiol. Plant Pathol. 13:253-258.

Weststeijn, E.A. 1981. Lesion growth and virus localization in leaves of *Nicotiana tabacum* cv. Xanthi nc. after inoculation with tobacco mosaic virus and incubation alternately at 22° C and 32° C. Physiol. Plant Pathol. 18:357-369.

Whatley, M. H., Hunter, N., Cantrell, M.A., Hendrick, C.A., Sequeira, L., and Keestra, K. 1980. Lipopolysaccharide composition of the wilt pathogen, *Pseudomonas solanacearum*. Correlation with the hypersensitive response in tobacco. Plant Physiol. 65.

Wheeler, H. 1975. Plant Pathogenesis. Springer Verlag, New York.

White, N.H., and Baker, E.P. 1954. Host pathogen relations in powdery mildew of barley 1. Histology of tissue reactions. Phytopathology 44:657-662.

White, R.F. 1979. Acetylsalicylic acid (aspirin) induces resistance to tobacco mosaic virus in tobacco. Virology 99: 410-412.

Whitehead, J.M., Dey, P.M., and Dixon, R.A. 1982. Differential patterns of phytoalexin accumulation and enzyme induction in wounded and elicitor-treated tissues of *Phaseolus vulgaris*. Planta 154:156-164.

Williams, P.H., and Keen, N.T. 1967. Relation of cell permeability alterations to water congestion in cucumber angular leaf spot. Phytopathology 57: 1378-1385.

Williamson, R.E., and Ashley, C.C. 1982. Free Ca^{2+} and cytoplasmic streaming in the alga *Chara*. Nature 296:647-670.

Willis, D.K., Rich, J.J., and Hrabak, E.M. 1991. *hrp* genes of phytopathogenic bacteria. Mol. Plant-Microbe Interact. 4:132-139.

Winans, S.C., Kerstetter, R.A., and Nester, E.W. 1988. Transcriptional regulation of the *virA* and *virG* genes of *Agrobacterium tumefaciens*. J. Bacteriol. 170:4047-4054.

Wingate, V.P.M., Norman, P.M., Lamb, C.J. 1990. Analysis of the cell surface of *Pseudomonas syringae* pv. *glycinea* with monoclonal antibodies. Molec. Plant-Microbe Interact. 3:408-416.

Wittmann, H.G., and Wittmann-Liebold, B. 1966. Protein chemical studies of two RNA viruses and their mutants. Cold Spring Harbor Symp. Quant. Biol. 31:163-172.

Wolpert, T.J., Macko, V., Acklin, W., Jaun, B., Seibi, J., Meili, J., and Arigoni, D. 1985. Structure of victorin C, the major host-selective toxin from *Cochliobolus victoriae*. Experientia 41: 1524-1529.

Wood, R.K.S. 1967. Physiological plant pathology. Blackwell Scientific Publications, Oxford.

Woods, A.M., Fagg, J., and Mansfield, J.W. 1988. Fungal development and irreversible membrane damage in cells of *Lactucae sativa* undergoing the hypersensitive reaction to the downy mildew fungus *Bremia lactucae*. Physiol. Mol. Plant Pathol. 32:483-497.

Wu, J.H. 1973. Wound-healing as a factor in limiting the size of lesions in *Nicotiana glutinosa* leaves infected by the very mild strain of tobacco mosaic virus (TMV-VM). Virology 51:474-484.

Wurzburger, J., Frimer, A.A., and Leshem, Y.Y. 1984. Inhibition of oxidative catabolism and oxy free radical production as a possible mode of cytokinen control of foliar senescence. Plant Growth Regul. 2:143-146.

Xiao, Y., Lu, Y., Heu, S., and Hutcheson, S.W. 1992. Organization and environmental regulation of the *Pseudomonas syringae* pv. *syringae* 61 *hrp* cluster. J. Bacteriol. 174:1734-1741.

Yamamoto, H., and Tani, T. 1986. Possible involvement of lipoxygenase in the mechanism of resistance of oats to *Puccinia coronata avenae*. Phytopathology 116:329-337

Yamayuchi, A., and Hirai, T. 1959. The effect of local infection with tobacco mosaic virus on respiration of leaves bearing local lesions. Phytopathology 49:447-449.

Yoshikawa, M., Keen, N.T., and Wang, M.C. 1983. A receptor on soybean membranes for a fungal elicitor of phytoalexin accumulation. Plant Physiol. 73:497-506.

Yoshikawa, M., and Masago, H. 1977. Activated synthesis of poly(A)-containing messenger RNA in soybean hypocotyls inoculated with *Phytophthora megasperma* var *sojae*. Physiol. Plant Pathol. 10:125-138.

Yoshikawa, M., Yamauchi, K., and Masago, H. 1978. Glyceollin: its role in restricting fungal growth in resistant soybean hypocotyls infected with *Phytophthora megasperma* var. *sojae*. Physiol. Plant Pathol. 12:73-82.

Young, J.M. 1974a. Development of bacterial populations in vivo in relation to plant pathogenicity. N. Z. J. Agric. Res. 17:105-113.

Young, J.M. 1974b. Effect of water on bacterial multiplication in plant tissue. N. Z. J. Agric. Res. 17:115-119.

Yucel, I., Xiao, Y., Hutcheson, S.W. 1989. Influence of *Pseudomonas syringae* culture conditions on initiation of the hypersensitive response of cultured tobacco cells. Appl. Environ. Microbiol. 55:1724-1729.

Zambryski, P. 1988. Basic process underlying Agrobacterium-mediated DNA transfer to plant cells. Ann. Rev. Genet. 22:1-30.

Zimmer, D.E. 1970. Fine structure of *Puccinia carthami* and ultrastructural nature of exclusionary seedling-rust resistance of safflower. Phytopathology 60:1157-1163.

Zimmer, D.E., and Schaffer, J.F. 1961. Relation of temperature to reaction type of *Puccinia coronata* on certain oat varieties. Phytopathology 51:202-203.

Zimmermann, U. and Steudle, E. 1974. Hydraulic conductivity and volumetric elastic modulus in giant algal cells: pressure-and volume-dependence. Pages 64-78 in: Membrane Transport in Plants. U. Zimmermann, and J. Dainty, eds. Springer Verlag, New York, Heidelberg, Berlin.

Zook, M.N., Rush, J.S., and Kuc, J.A. 1987. The role for Ca^{2+} in the elicitation of rishitin and lubimin accumulation in potato tuber tissue. Plant Physiol. 84:520-525.

Zucker, M., and Hankin, L. 1970. Physiological basis for a cycloheximide-induced soft rot of potatoes by *Pseudomonas fluorescens*. Ann. Bot. (Lond.) 34:1047-1062.

INDEX

About the Authors

Professor Anton J. Novacky obtained his education and early professional training at Charles University, Prague and Comenius University, Bratislava. He held postdoctoral fellowships with Professor B. A. Rubin, University of Moscow, Russia, and Professor Harry Wheeler, University of Kentucky and joined the Department of Plant Pathology, University of Missouri in 1969. Dr. Novacky's research has been concerned with alterations of plant cell membranes during the development of both pathogenesis and the hypersensitive defense reaction. Seeking more sensitive procedures to monitor cell membrane perturbations, he developed and introduced electrophysiological methods to measure transmembrane electropotentials in tissues and single cells of higher plants. With these techniques he has investigated the effect of fungal host-selective toxins on host cell membranes and cellular alterations elicited by bacteria during the development of the hypersensitive reaction. Additional research and teaching activities have been focused on primary and secondary active transport in plants. Together with his students and research associates he has published numerous journal articles and chapters in books. Professor Novacky has kept in close contact with his colleagues in Bratislava, and following the velvet revolution he was able to return to his Alma Mater, Comenius University on several occasions. He has encouraged several young scientists from Comenius to join him in Columbia, Missouri. At this writing he is a Fulbright Fellow in Bratislava teaching courses in plant physiology and the physiology of plant disease resistance.

Professor Robert N. Goodman obtained his academic degrees at the University of New Hampshire and the University of Missouri. He joined the Department of Horticulture, University of Missouri as a Post Doctoral Fellow in 1952 and the regular faculty in that same year. Dr. Goodman remained in the Department of Horticulture until 1967 when the Department of Plant Pathology was formed at Missouri. In 1968 Dr. Goodman became chairman of the Department and held that post through 1979. He held the position of professor until his retirement in 1991. During sabbatical leaves Professor Goodman worked in the laboratories of Professor Ernst Gäumann ETH, Zurich, Professor R. D. Preston, University of Leeds, Professor Nathan Sharon, Weizmann Inst., Israel and at Unilever Laboratories, Bedford, U.K. Most of Dr. Goodman's research dealt with plant pathogenic bacteria: their control with antibiotics, ultrastructure, infection process and the plants' defenses against them including HR, a subject into which he was indoctrinated by Professor Zoltan Klement, Budapest. With his students and research associates Professor Goodman published more than 100 journal articles, several chapters in books and two advanced treatises on The Biochemistry and Physiology of Plant Disease. The first was written with Professors Zoltan Kiraly, Budapest, Hungary, and Milton Zaitlin, Cornell University; the most recent edition with Professor Kiraly and Professor K. R. Wood, University of Birmingham, U.K. During the last decade, Professor Goodman has concentrated his research efforts on deriving remedial procedures for the control of *Agrobacterium tumefaciens* in grape. In retirement he is an adjunct research professor at the Missouri Fruit Experiment Station, Mtn. Grove, MO. In this way he is able to monitor his family's commercial vineyard and continue to conduct relevant research.

R. N. Goodman (left) and A. J. Novacky (right). Photograph by Roberto Celli, Director, Rockefeller Study Center, Villa Serbelloni, Bellagio, Italy.